QUALITY CONTROL

Principles and practice in the microbiology laboratory

QUALITY CONTROL

Principles and practice
in the microbiology laboratory

Edited by
JJS Snell
ID Farrell
C Roberts

PUBLIC HEALTH LABORATORY SERVICE

Public Health Laboratory Service
61 Colindale Avenue
London NW9 5DF

© Public Health Laboratory Service 1991. All rights reserved. No part of this book may be reproduced in any form or by any means without the written permission of the publisher.

ISBN 0 901144 31 2

Typeset by The Laverham Press, Salisbury, Wiltshire
Printed and bound in England by Hollen Street Press

Acknowledgements

The editors gratefully acknowledge the help of the following in reviewing and editing some of the chapters: Joan R Davies, former Deputy Director (retired), Public Health Laboratory Service; TH Flewett, Regional Virus Laboratory, Birmingham; and RJ Gilbert, Food Hygiene Laboratory, Central Public Health Laboratory.

Editors

JJS Snell is Director, Quality Assurance Laboratory, Central Public Health Laboratory
ID Farrell is Director, Birmingham Public Health Laboratory
C Roberts is Deputy Director, Public Health Laboratory Service

Foreword by Sir Joseph Smith, Director, Public Health Laboratory Service

Contributors

JA Bird	Microbiology Section, Unilever Research
DFJ Brown	Cambridge Public Health Laboratory
EO Caul	Bristol Public Health Laboratory
PL Chiodini	Department of Parasitology, London Hospital for Tropical Diseases
A Curry	Manchester Public Health Laboratory
JM Darville	Bristol Public Health Laboratory and University of Bristol
MH Greenwood	Poole Public Health Laboratory
TG Harrison	Division of Microbiological Reagents, Central Public Health Laboratory
WL Hooper	Poole Public Health Laboratory
DWR Mackenzie	Mycology Reference Laboratory, Central Public Health Laboratory
A Malic	Novo Nordisk Diagnostics Ltd
RJ Martin	Unipath Ltd
AH Moody	Department of Parasitology, London Hospital for Tropical Diseases
L Odwell	Department of Parasitology, London Hospital for Tropical Diseases
KD Phillips	(Formerly) PHLS Anaerobe Reference Unit, Luton Public Health Laboratory
DS Reeves	Department of Medical Microbiology, Southmead Hospital
JJS Snell	Quality Assurance Laboratory, Central Public Health Laboratory
LO White	Department of Medical Microbiology, Southmead Hospital
DJ Wood	National Institute for Biological Standards and Control
AE Wright	Former Director (retired), Newcastle Public Health Laboratory

Foreword

Quality control and assessment are essential aspects of laboratory medicine. They have a number of objectives – for example, contributing to cost-effective laboratory management – but the first aim is to eliminate preventable inaccuracy and also unnecessary delay. The consequences of an incorrect laboratory report in microbiology, as in any other branch of pathology, are potentially serious for the patient: a false-positive or false-negative result in an AIDS test, or its attribution to the wrong patient, could have devastating consequences for the individual. Incorrect results could also be serious for the public health: prevention, early detection or control of an outbreak can depend upon rapid, accurate laboratory tests. For example, incorrect or delayed identification of cases of typhoid could preclude early tracing and control of their source and be instrumental in leading to a large outbreak.

Quality control procedures also contribute to satisfactory standardisation of results between different laboratories, or of results in one laboratory over different time periods: failure to standardise makes the clinical interpretation of laboratory findings more difficult and reduces or may even negate their epidemiological value.

The need for quality control and assessment is underlined by the very nature of the science of microbiology, concerned as it is with living organisms which often display a great capacity for genetic variation. Detection and identification of microbes involves the use of potentially variable reagents such as antisera, or of inherently variable growth media, including living tissue culture cells. In consequence, many of the techniques of microbiology require the use of biological reference preparations or standards in order to provide acceptable reliability and reproducibility of test procedures.

The medical microbiologist's day-to-day task of interpreting and reporting laboratory results requires an appreciation – which can only come from the application of quality control – of the confidence that can be placed in the various procedures in use in that laboratory, including the means by which the specimens were transported to the laboratory, the sensitivity and specificity of the test methods and reagents used, and their reproducibility in day-to-day practice. Indeed, the very reporting procedure itself has to be quality controlled, not least to ensure minimum delay in the report reaching the

Foreword

intended recipient and that the intended information is correctly conveyed.

The need for a sound knowledge of the principles and practice of microbiological quality control is therefore well understood by medical microbiologists. Laboratory staff operate internal quality control procedures and take part in external quality assessment schemes. In the United Kingdom, the Public Health Laboratory Service (PHLS) is responsible for providing, on behalf of the Department of Health, the UK National External Quality Assessment Scheme (UKNEQAS) for microbiology in which PHLS, National Health Service and private laboratories, as well as laboratories abroad, participate to help ensure that they are operating to the highest possible standards.

This manual presents the principles of quality control and their application to the various aspects of laboratory practice in medical microbiology. The authors, from the PHLS, the National Health Service and commercial life, are expert in their various fields and their contributions will undoubtedly be of great value in maintaining the high quality of medical microbiology in the United Kingdom.

Sir Joseph Smith

Contents

Introduction: general aspects of quality control JJS Snell		13
1	*External quality assessment* JJS Snell	18
2	*Culture media* RJ Martin	24
3	*Bacteriological characterization tests* JJS Snell	37
4	*Antibiotic susceptibility testing* DFJ Brown	47
5	*Antibiotic assays* LO White and DS Reeves	63
6	*Anaerobic bacteriology* KD Phillips	77
7	*Preservation of control strains* JJS Snell	87
8	*Immunoassays* TG Harrison and A Malic	95
9	*Virus isolation* JM Darville and EO Caul	111
10	*Electron microscopy* A Curry and DJ Wood	124
11	*Mycology* DWR Mackenzie	138
12	*Parasitology* AH Moody, L Odwell and PL Chiodini	142
13	*Water microbiology* AE Wright	147

Contents

14 *Food microbiology – a PHLS perspective*   153
    MH Greenwood and WL Hooper

15 *Food microbiology – an industrial perspective*   164
    JA Bird

Introduction: general aspects of quality control

JJS Snell

The head of a department is responsible for the quality of the work carried out in the department. This is controlled by in house quality control procedures, the efficacy of which can be checked by participation in external quality assessment schemes.

Quality control is a continual monitoring of working practices, equipment and reagents and is the means by which the quality of microbiology services can be assured. The need for quality control is not limited to technical procedures undertaken in the laboratory. Factors which inter-relate with technical procedures in the chain of patient care are shown in Table 1. As errors can occur at any step in the chain, quality control is needed at all stages to minimize the

Table 1 The chain of patient care

Patient
Clinician orders test
Clerical work
Collection and labelling of specimen
Transport of specimen to laboratory
Clerical work
Technical analysis
Interpretation by microbiologist
Clerical work
Report to clinician
Interpretation by clinician
Diagnosis and treatment

occurrence of such errors. Thus, for example, there is little point in using a precise and accurate method for determining the number of microorganisms in urine if the sample is badly taken and left standing above a radiator for several hours before despatch to the laboratory, or similarly mistreated after it arrives in the laboratory.

Details of quality control procedures specific to particular areas of microbiology are given in the chapters of this manual. Their success relies on the presence of a basic infrastructure of good laboratory practice.

Good laboratory practice

This area covers all the organizational aspects that provide the infrastructure for the well-organized laboratory. It includes such diverse areas as staff training, provision of methods manuals and checking of reagents. Examples of this are calibration of instruments, monitoring of refrigerator, waterbath, oven and incubator temperatures, preventive maintenance of equipment and control of autoclaves. Failure to control these aspects is characteristic of what has been termed 'the undisciplined laboratory'[1] and has been discussed in detail in several publications.[2,3,4] Procedures and schedules for the monitoring and maintenance of general laboratory equipment will vary according to the type and sophistication of the equipment and frequency of use. Central to any control scheme is a system of documentation, which details the monitoring procedures to be performed, states when they should be performed, records the results of the control procedure, describes acceptable limits of results for the control procedure, describes the action required if results are outside these limits and records that this action has been taken. Documentation must also give details of servicing and maintenance procedures required and allow logging of these activities.

A basic factor influencing the design of a quality control programme is whether in house or commercially-produced reagents are in use. This is not the place for a discussion of the merits of either approach, but it is probably true that the use of commercially-produced reagents increases the likelihood that different laboratories will get the same results whether right or wrong. The trend towards the use of commercially-produced reagents is pronounced in the case of culture media, identification kits and serological tests. The use of 'home-made' reagents is frequently justified on the grounds that their

Introduction

cost is lower than that of commercially-produced reagents. A factor often ignored is that part of the increased cost is a consequence of the quality control carried out by the manufacturer, which relieves the user of some of this chore. Reputable manufacturers have placed increasing importance on quality control in recent years and the sophistication and extent of some of the procedures, such as the use of melting temperature analysis or high pressure liquid chromatography to determine the purity of chemicals, would be very difficult to emulate in most laboratories. That such basic quality control may be necessary is illustrated by the finding in the author's laboratory that some batches of maltose contain sufficient amounts of glucose to give false-positive results in fermentation tests. Since purchasers of commercially-produced reagents have paid for an implied guarantee of quality, they should insist that specifications are consistently met.

A decision must be made as to the necessity for and frequency of testing of commercially-produced reagents. This decision can only be made with a knowledge of the incidence of batch variation. The manufacturers must accept the primary responsibility for the quality of their product and if batch variation requires frequent in house quality control then a change of supplier is indicated. Unnecessary quality control testing is wasteful.[5] However, any decisions to reduce the level of routine in house quality control must be based on the proven reliability of a product rather than on an unproven assumption that all is well.

There is often confusion between product evaluation and quality control: the former measures what a product is capable of and defines its operating characteristics; the latter simply shows that batches of the product are meeting pre-defined standards. The distinction is important, as a product may be doing precisely what it is designed to do, but not what the user wants it to do. Another confounding factor is the tendency for users to modify methods by reducing volumes of reagents used for reasons of economy, by the use of in house produced reagents in combination with kit components, or by modification of procedures. Such modifications are not necessarily harmful but they are quite likely to alter the operating characteristics of the test, and most laboratories have neither the time nor facilities to determine the effects. It is important to appreciate that modifications of such methods may have consequences that are difficult to predict.

Quality control in the microbiology laboratory

Provision of a manual of methods

The provision of a methods manual is of fundamental importance in establishing good laboratory practice. Methods handed on from worker to worker by word of mouth are subject to 'drift'. Thus, after a few years the methods used on the bench differ considerably from those used originally. This is particularly true if new entrants to a laboratory are taught procedures by junior employees; for this reason, such teaching should be undertaken by senior staff. Written methods sheets also do much to prevent this drift. It is not sufficient simply to refer to a textbook for details of a method, as the actual details of preparation in the laboratory are important.

There is no need for a methods manual to be a single bulky document, and indeed there are some advantages in splitting it into convenient sections. The format of the manual can be flexible, so, for example, card index files giving details of media preparation can be considered as part of the manual.

The manual should cover all procedures a laboratory undertakes as well as peripheral activities, such as specimen collection. Topics covered should include the following:

- specimens required for each requested investigation and instructions for collection
- laboratory tests to be routinely performed for each type of request
- test methods, details of media and reagent preparation, test procedures, the controls to be included, expected control results, interpretation and recording of results (see page 38 for suggested details)
- monitoring procedures and maintenance schedules for equipment.

To be useful, methods manuals should fulfil the following criteria.

They should be authoritative They should be written by a senior member of staff who has had practical experience of the methods. Writing a methods manual is a good opportunity for a general review of techniques and is also an excellent educational exercise for the person writing it.

They should be realistic Written instructions must relate to what is expected in practice rather than to some theoretical ideal: workers will quickly lose respect for a manual that contains impracticable instructions.

They should be up to date Written instructions should reflect current practice and the manual amended as necessary and revised yearly. Changes in a method may be beneficial but they should always be made with the approval of senior staff and made official by amending the manual accordingly.

They should be available One or more complete sets of the manual should be available for reference. Individual sections should be separated for convenient reference and kept on the relevant benches for day-to-day use. These bench copies can be enclosed in transparent plastic sheets to avoid wear and tear.

They should be used It is the responsibility of senior staff to ensure that the manual is actually used and not relegated to the shelf, to be brought forth only to impress visitors.

References

1 Fodor AR. Microbiological reagents in search of precision. *Hlth Lab Sci* 1968; **5**: 5–11.
2 Blazevic DJ, Hall CT, Wilson ME. Practical quality control procedures for the clinical microbiology laboratory. Cumitech no 3. Washington: American Society for Microbiology, 1976.
3 Bartlett RC, Allen VD, Blazevic DJ *et al*. Clinical microbiology. In: Inhorn SL, ed: Quality assurance practices for health laboratories. Washington: American Public Health Association, 1978, 871–1005.
4 Anon. Standard for Quality Assurance. Part 2. Internal quality control in microbiology. Beckenham: European Committee for Clinical Laboratory Standards 1984.
5 Bartlett RC, Rutz CA, Konopacki N. Cost effectiveness of quality control in bacteriology. *Am J Clin Pathol* 1982; **77**: 184–90.

1 External quality assessment

JJS Snell

Quality assessment is a system in which specimens of known but undisclosed content are introduced into the laboratory and examined by the same staff using the same procedures as would normally examine patients' specimens of the same type.

Quality assessment may be organized within the laboratory or externally. When organized internally, specimens of known content are introduced into the routine system by senior staff who receive and evaluate the reports.[1] In an external scheme, specimens of known content are sent to participating laboratories for examination and the results are reported to the organizing laboratory and evaluated. Both types of scheme act as an indicator of the effectiveness of internal quality control procedures. An internally-organized scheme has the advantage that specimens can be more easily disguised as part of the routine work and perhaps provides a more realistic appraisal of the average standard of work. Specimens from an external scheme are more difficult to camouflage and there is a temptation to give them special treatment. However, an externally-organized scheme has several advantages, including the provision of a wide variety of organisms, stable specimens of known characteristics, availability of repeat samples in case of failure and the chance to compare individual performance with that of the general standard of other participants.

Assessment and control

It is important to appreciate the distinction between quality control and quality assessment, the latter acting as a check on the former. It

External quality assessment

follows from this that quality assessment cannot be used as a substitute for quality control and that there is little point in examining external quality assessment specimens unless an in house quality control scheme is in operation. Failure to appreciate this may result in the effort that should be directed towards quality control being misdirected, with undue attention being paid to quality assessment specimens. This can be dangerous, since in order to safeguard the welfare of patients, the laboratory manager needs to be aware of the standard of routine work. Quality assessment specimens can provide this information only if they are treated as nearly as possible in the same way as patients' specimens. This has been clearly illustrated by Lamotte *et al*[2] who distributed split samples of sera containing drugs of abuse. Laboratories unknowingly received duplicate samples of each sera, one as part of a quality assessment programme and the other as a 'blind' specimen submitted through a hospital or clinic. Some data from this trial are shown in Table 1.1: the success rate was much higher with the samples clearly identifiable as quality assessment specimens than with the blind samples. It is evident that if special attention is given to quality assessment samples, the laboratory management may be unable to gain an insight into their routine standard of performance.

Although external quality assessment in microbiology has been slower to develop than in clinical chemistry, a number of national and regional schemes now exist in the United States of America, Belgium, France, Norway, Australia, Canada and the United Kingdom. External quality assessment is best carried out on a national basis rather

Table 1.1 Performance with blind and proficiency testing samples

	% of positive samples in which drug was found	
	proficiency testing	blind
Amphetamine	100	48
Barbiturate	100	72
Morphine	99	63
Methadone	98	80

Data taken from LaMotte *et al*[2]

than internationally, since the microbes encountered, reporting and referral practices, methods and antimicrobials used will vary from country to country.

The United Kingdom National External Quality Assessment Scheme for Microbiology (UKNEQAS) meets the standards for external quality assessment schemes suggested by the World Health Organization.[3] This scheme is endorsed, and in part funded, by the UK Department of Health (DoH) and is organized by the Public Health Laboratory Service (PHLS) from the Quality Assurance Laboratory (QAL) at the Central Public Health Laboratory, Colindale. The scheme is comprehensive and in addition to bacteriology and antimicrobial susceptibility testing, other categories of specimens regularly supplied include virus isolation and identification, rubella IgG serology, rubella IgM serology, general virus serology, hepatitis B serology, electron microscopy, HIV serology, chlamydia detection, blood parasitology, faecal parasitology, mycology, syphilis serology, slides for detection of acid and alcohol-fast bacilli and assay of antibiotics in serum.

Organization of the scheme

The working principles of the scheme have already been described.[4,5,6] In order to maintain confidentiality, participants are given a unique code number when they join the scheme by which they are identified in all routine transactions. Simulated clinical specimens are prepared in QAL and distributed to participants with request/report forms. Participants examine the specimens in their laboratories and report their findings to QAL. Reports are analyzed and participants receive a summary of the overall results for the distribution and a computer-derived analysis of their individual results on current and recent specimens. Participation is available to all health care microbiology laboratories in the UK, of which 514 currently participate.

The great majority of specimens are straightforward and reflect those likely to be regularly found in UK clinical practice. Occasionally, more unusual specimens may be distributed for educational purposes or where recognition of an unusual pathogen, such as *Corynebacterium diphtheriae* or *Vibrio cholerae*, may be of great importance to the patient or community. The proportion of positive speci-

External quality assessment

mens is of course higher than that found in routine practice. Details of specimen preparation have been previously described.[7,8]

The specimens are subjected to rigorous quality control in QAL and numerous samples are examined before, during and after distribution, including duplicate sets of specimens returned through the post from various geographically-dispersed laboratories. Where quality control checks during the distribution period reveal unexpected and unacceptable changes in the specimen, participants are informed of the problem and laboratories' scores for these specimens are not included in the calculation of performance statistics.

Any specimen issued as part of the scheme may contain fully virulent pathogens (other than those in hazard group 4[9]). In this respect they are identical to clinical specimens and participants are warned that they must be treated with due care and attention to safety procedures.

Three weeks are allowed between despatch of specimens and return of reports for most distribution types. After this period, repeat specimens are available to laboratories failing with the original.

To assist laboratories in evaluating their performance, a scoring scheme is used. Details vary according to the distribution type, but generally scores of 2 are awarded for correct results, 1 for partially correct results, 0 for wrong results and failure to return reports and −1 for grossly misleading results. Participants receive an individual computer print-out showing the scores awarded for their results with specimens included in the current distribution and an analysis of their performance over a six-month period. The information presented includes a list of the specimens supplied during the six months, the number of specimens reported as not examined by the laboratory, the number of reports not returned to QAL and the number of results received too late for analysis. Performance data is presented as the total score achieved by the laboratory (derived by adding together the scores for each specimen), the total possible score that would have been achieved with a fully correct result in all the specimens, and the average score for the series. The average score for a single specimen is the sum of the scores of all laboratories divided by the number of laboratories reporting. The average score for the series is the sum of the average scores of the individual specimens.

With the above information each laboratory can compare its performance with that of its peers. If a laboratory's total score is higher than the average score they are performing better than average. If their total score is lower than the average score their performance is below average. To enable laboratories to quantify their performance, the number of standard errors the individual's result is above or below the average score is calculated and appears on the print-out. This is simply a statistical device to help identify possible poor performance, which is defined as a total score more than 1.96 standard errors below the average score.

Participants should use the data provided on the computer print-outs to monitor their own performance and to take any action required as problems are revealed. An advisory panel for microbiology, comprised of experienced microbiologists nominated by the professional societies, exists to offer confidential help and advice to participants experiencing continuing problems. The confidentiality of results is very strictly maintained and the scheme organizer never reveals details of the performance of individual participants to individuals or organizations other than to the advisory panel in cases of persistent poor performance. This confidentiality is necessary to encourage participants to treat QA specimens in the same way as patients' specimens, thus allowing them to gain an insight into their routine capabilities.

Results

Results obtained by participants have been documented.[4,8,10] The general standard of performance of UK laboratories is, on the whole, satisfactory, with around 95 per cent of laboratories achieving full scores in any one specimen containing a common pathogen.

Most participants appear to find the scheme useful and good relationships have been maintained, because the scheme is confidential and is seen as being an educational service rather than a regulatory tool. Any UK clinical laboratories not currently participating in the scheme are urged to contact the author to obtain information on how to join. Participation is also available to a limited number of overseas laboratories at the discretion of the organizer.

References

1 Estevez EG. A program for in-house proficiency testing in clinical microbiology. *Am J Med Technol* 1980; **46**: 102–5.
2 LaMotte LC, Guerrant GO, Lewis DS, Hall CH. Comparison of laboratory performance with blind and mail-distributed proficiency testing

samples. *Public Health Rep* 1977; **99**: 554–60.

3 Report on a WHO working group: External quality assessment of health laboratories. World Health Organization, Regional Office for Europe, 1981.

4 Snell JJS, DeMello JV, Gardner PS. The United Kingdom national microbiological quality assessment scheme. *J Clin Pathol* 1982; **35**: 82–93.

5 Snell JJS. United Kingdom national external quality assessment scheme for microbiology. *Eur J Clin Microbiol* 1984; **4**: 464–7.

6 Reed SE, Gardner PS, Snell JJS, Chai, O. United Kingdom scheme for external quality assessment in virology. Part I. General method of operation. *J Clin Pathol 1980*; **38**: 534–41.

7 DeMello JV, Snell JJS. Preparation of simulated clinical material for bacteriological examination. *J Appl Bacteriol* 1985; **59**: 421–36.

8 Reed SE, Gardner PS, Stanton J. United Kingdom scheme for external quality assessment in virology. Part II. Specimen distribution, performance assessment, and analyses of participants' methods in detection of rubella antibody, hepatitis B markers, general virus serology, virus identification and electron microscopy. *J Clin Pathol* 1985; **38**: 542–53.

9 Anon. Advisory committee on dangerous pathogens: Categorisation of pathogens according to hazard and categories of containment. London: Her Majesty's Stationery Office 1984, 4–8.

10 Snell JJS, DeMello JV, Phua TJ. Errors in bacteriological technique: results from the United Kingdom national external quality assessment scheme for microbiology. *Med Lab Sci* 1986; **43**: 344–55.

2 Culture media

RJ Martin

Culture media are vital to microbiology: without good media there is little chance that good results will emerge from the laboratory.

Most culture media used in the northern hemisphere are commercially produced and will have passed quite extensive process and end-product control testing before release.

A brief summary of the manufacturing protocol that the culture medium has undergone before reaching the laboratory is as follows.

1   Raw materials – selected, tested or purchased to quality specifications.
2   Manufacturing – standard operating procedures, in-process controls, good manufacturing practice operations.
3   Quality tests – chemical and biological parameters checked to ensure end-products meet product quality specification; packaging, labelling and storage are important.

It would not be reasonable for the end-user to repeat all these tests. Neither would it be reasonable, however, to use culture media without quality control (QC) tests. As the medium undergoes reconstitution, heat-processing and perhaps supplementation with additives in the laboratory, it is essential to have controls over these processes.

The few laboratories which prepare media from basic raw materials must follow more comprehensive control procedures than outlined in this chapter and should refer to larger works on culture media preparation.[1,2]

Quality control is only part of total quality assurance,[3] which covers all aspects of good laboratory practice.[4] However, this chapter will

Culture media

concentrate solely on culture media testing in the laboratory.

It should be noted, however, that in spite of extensive quality testing, no culture medium is perfect, ie capable of recovering all strains of the desired organisms. Testing protocols are designed to measure the major or significant characteristics of the medium. Good results from these tests will give considerable assurance about the ultimate performance of the medium but cannot offer guarantees. All laboratories have experienced strains of organisms which have specific growth requirements not satisfied by standard media. It is important to be aware and vigilant about such variants or mutants.

Principles

The tests to be carried out in the laboratory on culture media should be cost-effective. At the simplest level this could involve a pH test and testing the growth of one or two organisms. Selective media should be tested with organisms which would be expected to grow and those which should be inhibited. QC tests will differ according to whether they are ongoing routine, selection of a new batch of medium or sampling from new suppliers. Each laboratory must decide on the amount and type of QC testing it requires. In this chapter a minimal QC test programme of culture media used in medical microbiological laboratories is described. Although this is a two-tier system of testing, the intensity of testing employed will depend on the circumstances. QC testing of the final product is not intended as a substitute for proper control of culture media processing but is a reliable adjunct and indicator of performance.

For each medium to be included in the test programme, a laboratory standard is selected from a known, acceptable batch and set aside for future use. Whenever required, the standard can be put through the normal processing and used as a comparison for the batch in use. The standard may only need to be used from time to time.

The two-tier QC testing programme will operate as follows:

1 new supplier/new product ⎱
2 supplier comparison   ⎰ extended panel of organisms

3 new batch of medium   extended or short as appropriate

4 routine batches made in ⎱
   laboratory or purchased ⎱ short panel of organisms.
   (eg ready poured plates) ⎰

Quality control in the microbiology laboratory

The test programme is based on the following basic parameters to be examined and recorded:

1 physical characteristics:

colour
clarity
gel strength } certain characteristics may not be indicated for specific media
pH
sterility

2 microbiological performance.

Physical characteristics must be noted because any change from normal may be a vital and easily recognized indicator. Thus a drift in pH may indicate a batch of medium has been over-autoclaved. More comprehensive reviews of sources of errors in preparation of media are readily available.[5,6,7]

Factors affecting the measurement of microbiological performance are:

- the size of the panel of organisms used (see above)
- choice of organisms.

Details of organisms recommended for the commonly-used culture media are shown in Tables 2.1 and 2.2.

Most test organisms have American Type Culture Collection (ATCC)/National Collection of Type Cultures (NCTC) numbers and are available from the normal sources.[8,9,10] Test strains have been selected as critical for each medium and are suitable indicators for routine monitoring of performance. The method of inoculation employed varies with individual choice. Some laboratories prefer to use the 'ecometric' method[11,12] but others will prefer the Miles Misra technique[13] or the classical inoculating technique with a wire loop. The organisms listed are intended to be used as the 'short panel' and as an 'extended panel' for occasional testing as described. Culture media manufacturers find that perception or expectation of a culture medium by laboratories varies so a laboratory may choose to add selected strains to meet its individual, local requirements.

Methods

Process control

Dust control Health and safety requirements and common sense indicate the need to control dust which may be distributed in the atmosphere during manipulations involving powders. The provision

Table 2.1 Quality control organisms

No		ATCC	NCTC
1	*Alcaligenes faecalis*	8750	11953
2	*Aspergillus niger*	16404	2275 (NCPF)
3	*Bacillus cereus*	10876	7464
4	*Bacillus subtilis*	6633	10400
5	*Bacteroides fragilis*	25285	9343
6	*Bacteroides melaninogenicus*	15930	11321
7	*Bordetella bronchiseptica*	4617	8344
8	*Bordetella parapertussis*	–	10521
9	*Bordetella pertussis*	–	8605
10	*Brucella abortus*	4315	–
11	*Campylobacter jejuni*	29428	11322
12	*Candida albicans*	10231	3179 (NCPF)
13	*Candida krusei*	–	3100 (NCPF)
14	*Citrobacter freundii*	8090	9750
15	*Clostridium difficile*	–	11204
16	*Clostridium histolyticum*	19401	503
17	*Clostridium perfringens*	13124	8237
18	*Clostridium sporogenes*	11437	–
19	*Corynebacterium diptheriae gravis*	19409	3984
20	*Corynebacterium diptheriae intermedius*	14779	10681
21	*Enterobacter aerogenes*	13048	10006
22	*Enterococcus faecalis*	29212	–
23	*Enterococcus faecalis*	19433	775
24	*Escherichia coli*	25922	10418
25	*Escherichia coli 0157: H7*	–	–
26	*Fusobacterium necrophorum*	–	10575
27	*Gardnerella vaginalis*	14018	10915
28	*Haemophilus influenzae*	19418	4560
29	*Klebsiella pneumoniae*	13883	9633
30	*Lactobacillus bulgaricus*	11842	–
31	*Legionella pneumophila*	33152	11192
32	*Listeria monocytogenes*	11994	–
33	*Mycoplasma hominis*	23114	10111
34	*Mycoplasma pneumoniae*	15531	10119
35	*Neisseria gonorrhoeae*	19424	8375
36	*Neisseria meningitidis*	13077	10025
37	*Penicillium notatum*	9178	–
38	*Peptococcus magnus*	15794	11804
39	*Peptostreptococcus anaerobius*	27337	11460
40	*Proteus mirabilis*	10975	10823 (NCIB)
41	*Proteus vulgaris*	13315	4175
42	*Pseudomonas aeruginosa*	27853	10662
43	*Pseudomonas cepacia*	17759	10661
44	*Salmonella derby*	–	5721
45	*Salmonella dublin*	–	9676
46	*Salmonella enteriditis*	13076	–
47	*Salmonella typhimurium*	14028	12023
48	*Salmonella virchow*	–	5742
49	*Serratia marcescens*	13880	10211
50	*Shigella flexneri*	12022	–
51	*Shigella sonnei*	25931	–
52	*Staphylococcus aureus*	25923	6571
53	*Staphylococcus aureus*	6538	7447
54	*Staphylococcus epidermidis*	12228	–
55	*Streptococcus agalactiae*	13813	8181
56	*Streptococcus pneumoniae*	6303	–
57	*Streptococcus pyogenes*	19615	–
58	*Streptococcus thermophilus*	14485	–
59	*Streptococcus sp viridans type*	–	1080
60	*Vibrio furnissii*	–	11218
61	*Vibrio parahaemolyticus*	–	10885
62	*Yersinia enterocolitica*	–	10460
63	*Yersinia enterocolitica*	–	10462

Table 2.2

Culture medium	Short panel		Extended	
	+	−	+	−
Amies transport medium	35, 52		56, 57	
Andrade peptone water	test with specific sugars and organism			
Bacillus cereus selective agar	3, 4	24	17, 52 (different reactions)	42
Baird Parker medium	52	4	53	40, 54 (diff reaction)
Bismuth sulphite agar	46, 47	14, 24	44, 45, 48	
Blood agar base	52, 56, 57		28, 42, 59	
Blood agar base no 2	52, 56, 57		28, 42, 59	
with Brucella supplement	10	24	additional *Brucella* strains	22, 42, 52
with Campylobacter supplement	11	24	additional *Campylobacter* strains	23, 40
with Gardnerella supplement	27	29	additional *Gardnerella* strains	12, 42, 52
with Staph/Strept supplement	52, 57	24	23, 53	40, 42
with Strept supplement	57	52	22, 55	24, 40
Bordet–Gengou agar	9		7, 8	
Brain heart infusion agar	36, 56		52, 57	
Brain heart infusion broth	36, 56		52, 57	
Brilliant green agar	45, 46	24, 41	47, 48	
Brilliant green bile 2% broth	21, 24	4, 52	24 (Gas 44°C)	21 (no gas 44°C)
Brucella agar base	10			
with Brucella supplement	10	24		23, 42, 52
Buffered peptone water	47			
Campylobacter medium	11	24	additional *Campylobacter* strains	22, 40
Charcoal agar	28		7, 8, 9	
with Cephalexin	8, 9	29, 52		
CLED	40, 52		22, 24, 42	

Culture media

Medium				
Clostridium difficile agar	15	24		17, 52
Cooked meat medium	{ 16 (Proteolysis) { 17 (Saccharolysis)	4, 18		
Corn meal agar	12	13		
Desoxycholate citrate agar	47, 51	24	46, 50	40
Desoxycholate citrate lactose sucrose agar	47, 51	24	46, 50	40
Diagnostic sensitivity test agar			{ tested with selected antimicrobial discs and organisms { for zones of inhibition	
with Blood	36, 56		36, 56	
without Blood	23, 42, 52		53	
Dnase agar	52	54		
Gardnerella vaginalis selective medium	27	29	additional *Gardnerella* strains	12, 42, 52
GC agar base				
with Antibiotics	35, 36	41, 52	additional *N gonorrhoeae* strains	
without Antibiotics	35			
Hartley's digest broth	56, 57			
Hektoen enteric agar	47, 51	24	46, 50	40
Hoyle medium	20	22	19	
Isosensitest agar			{ test with selected antimicrobial discs and organisms for zones { of inhibition	
with Blood	36, 56		test MIC of selected organisms	
without Blood	22, 42, 52			
Isosensitest broth	22, 52, 57			
Kranep agar	52	54	53	
Lauryl tryptose manritol broth with tryptophan	24	52	24 (Indole 37°C, 44°C)	22, 40
Legionella selective medium	31	24, 54	additional *Legionella* strains	22, 42
Listeria selective agar base	32	52	additional *Listeria* strains	22, 24
L–S differential medium	58	30		
Litmus milk	1 (alkaline pH) 17 (Alkaline pH) (stormy clot) 42 (peptonization)		24, 52	

Table 2.2 (*cont.*)

Culture medium	Short panel +	Short panel −	Extended +	Extended −
MacConkey agar	22, 52		24, 40, 42, 47	
MacConkey agar no 2	22	52	24, 40, 42, 47	
MacConkey agar no 3	24	22	40, 42, 47	52
MacConkey broth	24 (acid plus gas)	52	24 (gas 37° and 44°C)	21 (gas 37°C, no gas 44°C)
Malt extract agar	2, 12	3 (at pH 3.5)		
Mannitol salt agar	52	24	53, 54	40
McBride medium	32	52	additional *Listeria* strains	24
MCLB medium	47, 48	24	46	40
Membrane enriched teepol broth	21, 24		21, 24, 40 (37°C 44°C)	
Membrane lauryl sulphate broth	21, 24		21, 24, 40 (37°C 44°C)	
Minerals modified medium base	24 (acid plus gas)		24 (simulated coli count from water sample)	
Mueller Hinton agar	22, 42, 52		test with selected antimicrobial discs and organisms for zones of inhibition	
Mycoplasma agar	34	24	33	40, 41
Mycoplasma broth	34	24	33	40, 41
Nutrient agar	52, 57		42, 59	
Nutrient broth	52, 57		42, 59	
Peptone water	24 (Indole 37°C)		{ 24 (Indole 37°C, 44°C) { fermentation + of sugars	
Plate count agar	24, 52		{ compare batches of using seeded { milk counts	
Pseudomonas agar base	42	24	43 (growth dependent on supplement used)	41, 52

Rappaport broth	46	24	47, 48	
Sabouraud dextrose agar	12		2, 13, 37	
Salt meat broth	52		53	
Schaedler anaerobic agar	17		5, 52	
Schaedler anaerobic broth	17		5	
Selenite broth	46 (sub to Mac agar)		47, 48	
Simmons citrate agar	46	24	21, 47	
Slanetz and Bartley medium	22	24	23	42
Sorbitol MacConkey agar no 3	25	24		22, 41, 52 (NG)
SS agar	47, 51	24	46, 50	40
Stuarts transport medium	56 (recovery after three days)		19, 57 (recovery after 3 days)	
TCBS cholera medium	60	24	61	41
Thioglycollate medium	17, 52		4, 12, 18	
Tinsdale medium	20	22	19	52
Todd–Hewitt broth	57		22, 56, 59	
Tryptone bile agar	24 (Indole 44°C)			29
Tryptone soya agar	52, 56, 57		28 (with blood) 42, 59	
Tryptone soya broth	52, 56, 57		12, 42, 59	
Tryptose phosphate broth	22, 52		57, 59 (also used in tissue culture media)	
Violet red bile agar	21, 24	52	21, 24 (count in milk or water)	41, 47
Wilkins–Chalgren anaerobe agar	5, 17		18, 22	
NAT medium	6, 38	24		
NAV medium	5, 26	24		
Wilkins–Chalgren anaerobe broth	5, 17		18, 22	
XLD	47, 51	24	46, 50	40
Yeast extract agar	22, 52		55	
Yersinia selective agar	62	24	63	29, 40

Quality control in the microbiology laboratory

of gloves and suitable masks is a basic requirement for workers handling these materials.

Weighing Accurate weighing of powders on a suitable balance is required. Check the balance periodically.

Water quality High quality distilled or demineralized water is required. Check conductivity and pH.

Label instructions Follow instructions for preparation and use of a culture medium.

Heat processing Culture media may deteriorate and lose performance during heat processing and this will continue as the heating cycle is extended. It is therefore recommended that heating cycles are planned using thermocouples and well controlled: an ideal cycle is the minimum which destroys all organisms.

Batch rotation It is important that batches of media are used in batch number order.

Record keeping Keep adequate records of batches of medium received, processed, tested and used. It may be necessary to undertake an audit or investigation of batch failure, for which records are essential. Simple forms can be designed for this purpose.

Physical characteristics

Colour Compare colour of reconstituted and sterilized medium against the laboratory standard or the expected normal colour.

Clarity Compare medium to standard or expected clarity.

Gel Test a freshly poured and surface-dried plate with a wire loop; the agar medium should be firm and workable, but not over-hard. Over-solid gel often results in small colonies.

pH Test the pH in a beaker with a suitable pH electrode; gels can be tested at ambient temperature in a beaker or on a plate.

Culture media

Sterility Take one or two plates from a batch and incubate for 24–48 h at room temperature (18°C) and 37°C and examine for bacterial contamination.

Microbiological performance

Solid media

There are several alternative techniques which may be employed for routine QC of plates. Each has its merits, and workers will find that one particular method is more suitable in their laboratory. A mixture of methods may be indicated. Inocula must not be too heavy since this may give rise to atypical growth.

Any method Any method may be selected to give single isolated colonies.

Streaking by standard method Employ a rigidly standard method of streaking with a wire loop to give single colonies. This is a simple, practical and economic method to use which permits comparison between test and standard media.

Ecometric method[11,12] A plate is divided into sections and inoculum carried from one section into another by a standardized procedure.

Miles–Misra technique[13] *(modifications)* This method calls for dilutions of liquid cultures to be 'spotted' on to plates so that resultant numbers of colonies can be compared between test and standard. In one variation, the drops of bacterial specimens are permitted to dry without spreading, but an alternative way is to spread the drop by the application of a wire loop.

Liquid media

The assessment of performance of liquid media can often be more difficult than solid media. This fact particularly applies to selective media, eg, for salmonella. There is the perennial question of whether to test such media with pure cultures or mixtures of appropriate organisms. Each approach has its advantages and disadvantages and workers must choose their own QC protocol.

Sensitivity (susceptibility) testing media

These media have their own special characteristics. Generally, culture media manufacturers spend a great deal of time in selection of raw materials and testing of final product. However, laboratories should check new batches of medium with a small range of antibiotic discs against suitable control organisms, eg strains 24 and 52 (see Table 2.1). Sensitivity testing QC must also include routine surveillance of patterns of results.

Storage of test organisms

A quality control system requires availability of a number of cultures, probably held in two or more storage systems.

Original source   These are often freeze-dried ampoules from national or commercial collections.

Freeze-dried/liquid nitrogen   This is for long-term storage.

Plates/slopes   These are for day to day use. Resuscitating a new ampoule from freeze-dried/liquid nitrogen is necessary from time to time. It may be stored at room temperature for seven days prior to sub-culturing.

Some cultures will require special storage conditions, eg anaerobic organisms.

Interpretation

The advantage of including a laboratory standard is that interpretation of results is aided by the reference material. Where a laboratory standard is not employed, then an arbitrary performance standard has to be employed which requires development in each laboratory.

Solid media

Assessment of a medium is based on:
- colonial morphology, reactions etc
- quantitative comparison: for Miles–Misra technique a recovery of

Culture media

$\geqslant 70\%$ of number of colonies should indicate a satisfactory performance.

Liquid media

For the Miles–Misra technique, approximately equal growth (measured by eye) produced at similar inoculum levels will indicate satisfactory performance compared to the standard.

For selective media, sub-culture on to an appropriate solid medium enables critical comparison to be made. If mixtures of organisms are employed, it is preferable that sub-culture is made on to a medium which will reveal all test organisms, since a very selective medium (eg DCA) will obscure the true mixture. Further details of assessment of liquid and solid media are described by Baird et al.[12]

References

1 Bridson EY, Brecker A. Design and formulation of microbial culture media. In: Norris JR, Ribbons DW, eds: Methods in microbiology, vol 3A. London: Academic Press, 1970.

2 Bridson EY. Natural and synthetic culture media for bacteria. In: Rechcigal M Jr, ed: Handbook series in nutrition and food, vol III. Ohio: CRC Press, 1978.

3 Martin RJ. Quality assurance and clinical microbiology. *Med Lab Sci* 1983; **40**: 269–74.

4 Department of Health and Social Security. Good laboratory practice. The UK compliance programme. London: DHSS, 1986.

5 Barry AL, Fay GD. A review of some common sources of error in the preparation of agar media. *Amer J Med Technol* 1972; **38**: 241–5.

6 Martin RJ. Storage of microbial culture media. *Lab Practice* 1971; **20**: 653–6.

7 Oxoid Manual. Wade Road, Basingstoke, Hants RG24 0PW.

8 American Type Culture Collection. 12301 Parklawn Drive, Rockville, Maryland 20852 – 1776 USA.

9 National Collection of Type Cultures. Central Public Health Laboratory, 61 Colindale Avenue, London NW9 5HT.

10 DIFCO. Difco Laboratories, PO Box 331058, Detroit M48232 – 7058 USA.

11 Mossel DA, Van Rossem F, Koopmans M *et al*. Quality control of solid culture media: a comparison of the classic and the so-called ecometric techniques. *J Appl Bact* 1980; **49**: 439–54.

12 Baird RM, Corry JEL, Curtis GDW. Pharmacopoeia of culture media for food microbiology. *Int J Food Microbiol* 1987; **5**: 187–299.

13 Miles AA, Misra SS, Irwin JO. The estimation of the bacteriocidal power of the blood. *J Hyg* (Cambridge) 1938; **38**: 732–49.

Useful publications

Ellis RJ. Quality control procedures for microbiology laboratories, 3rd edn. Atlanta, Georgia: Department of Health and Human Services, Centers for Disease Control, 1981.

Bartlett RC *et al*. Quality assurance for commercially prepared microbiological culture media. Tentative standard M22-T. The National Council for Clinical Laboratory Standards 1987; 7: no. 5.

National Committee for Clinical Laboratory Standards. Performance standards for antimicrobial disk susceptibility tests, 4th edn. Tentative Standard M2–T4.

European Committee for Clinical Laboratory Standards. Standard for quality assurance; part 2; Internal Quality Control in Microbiology 1985; 2: no. 4.

Additional culture collections

National Collection of Pathogenic Fungi (NCPF)
Central Public Health Laboratory
61 Colindale Avenue
London NW9 5HT
UK

The Curator
National Collection of Industrial Bacteria (NCIB)
PO Box 31
135 Abbey Road
Aberdeen AB9 8DG
UK

3 Bacteriological characterization tests

JJS Snell

Lack of test reproducibility both within and between laboratories is a well-documented phenomenon.[1,2,3,4] The consequences affect not only taxonomic studies: as data from UKNEQAS has shown, substantial errors occur in the isolation and identification of medically-important pathogens from simulated clinical material.[5] This lack of test reproducibility may cause further variation between and within laboratories. Consistent differences in the results of tests between laboratories may be due to differences in the sensitivities of the methods used. The practical importance of such differences is that results given in diagnostic tables and keys may be applicable only if the stated methods are used. This is particularly true in the case of micro-methods or diagnostic kits. Unrecognized differences of this type lead to difficulties in the identification of strains and even misidentification if the test in question is the key one of a limited set. Thus, spurious production of acid from lactose may result in failure to recognize a strain of salmonella. The aim of quality control of characterization tests is to reduce to the minimum test variation within and between laboratories.

Although some consider the process of identification begins at the stage when a pure culture has been isolated, it starts in fact at the primary stages of microscopy and culture and includes the selection of colony types for further identification and the exclusion of those to be disregarded. Thus, the primary growth medium may profoundly affect subsequent stages in identification. Quality control of primary isolation media is discussed elsewhere in this book. However, criteria must be developed for the selection of organisms from the

Quality control in the microbiology laboratory

primary plate for further identification from each specimen type examined. Careful and accurate identification of multiple species isolated from a specimen may well be completely irrelevant to the patient's welfare.

As in all other areas of microbiology, quality control of characterization tests requires an infrastructure of good laboratory practice, as discussed in the introduction, and only quality control procedures directly relevant to characterization procedures are discussed here.

Standardization of test methods

Standardization of methods reduces variation both within and between laboratories. A great variety of methods and variations of methods are available for even the simplest tests.[6] An earlier UK-NEQAS survey revealed 21 methods or variations in use for testing motility alone! A laboratory may employ a particular method because of long experience and reliable results, or for no better reason than inertia. In the latter case, methods in use should be critically evaluated. If changes are considered desirable, then there is some virtue in choosing widely-used methods which have been shown to be reliable. The methods given by Cowan and Steel for the identification of medical bacteria[7] fulfil these criteria.

A basic decision that must be made is whether to use home-made or commercially-produced reagents; factors influencing such a choice are discussed in the introduction. It is evident that the use of commercial identification kits is now widespread in microbiology.

Details of test methodology may be critical. Both time and temperature of incubation and ingredients of media, such as the particular brand of peptone used and inoculum size, may affect the results of some tests. All these factors should be standardized within the laboratory.

Proper documentation of the tests and procedures used in a laboratory for characterization is necessary, and a chapter of the laboratory methods manual (see the Introduction) should be devoted to this subject. The chapter should contain the following information for each test:

A brief statement of the principle of the test   This will ensure that the staff concerned understand the test: the educated worker is probably the most important single factor in a quality control programme.

Bacteriological characterization tests

Formula of the medium This must be specific and include such details as the names and manufacturers of non-defined ingredients such as peptones; it must state if tap, distilled or deionized water is to be used and whether solutions of indicators are to be aqueous or, for example, alcoholic. The weight of salts containing variable water of crystallization should be adjusted to that appropriate to the substance actually used in the laboratory. The full chemical formula of the substance, including the water of crystallization, should also be stated. Quantities given should relate to normal batch sizes, but weights and volumes for likely alternative batch sizes should also be specified.

Details of preparation It is important that a senior member of staff has actually prepared the medium before writing the method, to uncover any problems and inconsistencies. Methods should be as simple as is consistent with a satisfactory product, to discourage the introduction of unofficial short cuts. Indicate what volume of container should be used to prepare the medium in. Always give a value for the pH of the medium and specify the normality of reagents used to adjust pH to avoid undue dilution of the medium. The stage at which pH is to be checked should be stipulated; it is often easier to check and adjust pH before addition of the agar or indicators. State the parameters for autoclaving in the same units as are actually indicated on the laboratory autoclave gauge, ie pressure or temperature.

Containers and volumes State the size of container used, the type of closure and the method of colour coding or labelling. The volume to be dispensed should be specified as this may be important: eg, gluconate broth dispensed in too large a volume will fail to heat up sufficiently on addition of Clinitest® tablets.

Description of appearance Describe the desired colour and opacity of the medium and state the limits of permitted deviation from this appearance.

Details of sterility testing State the number of plates or tubes to be tested, together with the temperature and length of incubation.

Preparation of reagent solutions Give the formula, reagent grade and method of preparation and storage conditions.

Quality control in the microbiology laboratory

Indication of shelf life   Give some estimate of the reliable shelf life of the medium.

Method of testing   This should include details of the inoculum to be used, method of inoculation, temperature, time and atmosphere of incubation.

Method of reading and recording   Give a clear description of the appearance of positive, weak positive and negative results, with a protocol for recording them. Failure to record results in a uniform manner may create confusion, especially with complex tests such as nitrate reduction or oxidation/fermentation.

Interpretation of results   Guidance on the interpretation of test results is needed so that unwarranted conclusions are not drawn from inadequate or misconstrued data. Experience from the UKNEQAS shows that common errors, which at first sight appear to indicate faults in reagents, often on closer examination prove to be caused by misinterpretation of results. This is particularly true in serology, where the complexity of antigenic schemata and the occurrence of cross-reactions present pitfalls for the unwary. Thus, although an experienced worker appreciates that organisms agglutinating with salmonella O sera are not necessarily salmonella, this may not be obvious to junior staff. Similarly, the presence of group-specific agglutinins left in some *Shigella flexneri* diagnostic antisera to facilitate the subtyping of certain serotypes can cause mistyping if interpretative criteria are not clearly laid down. The range of tests used must be adequate to support the level of identification attempted. Errors in which misidentification is probably due to an inadequate range of tests have repeatedly been shown in the UKNEQAS[5] for example, in the misidentification of *Moraxella phenylpyruvica* as *Neisseria gonorrhoeae*, of *N lactamica* as *N meningitidis* and of *Clostridium bifermentans* as *C perfringens*. It is better to report accurately to generic level than inaccurately to species level.

Suitable control strains   List specific strains giving positive and negative results in each test (see Tables 3.1 and 3.2), together with a schedule for testing.

References   Give literature references both to the original method and

Bacteriological characterization tests

to any subsequent modifications. Reprints of the relevant papers should be available in the laboratory.

Although the above list may appear formidable, many of the items can be dealt with concisely and the finished document need not be unduly bulky. The process of compiling such a detailed description for each medium encourages a systematic and logical approach and will almost certainly bring to light inconsistencies and areas where improvements can be made.

Stock control and shelf life

This is a difficult area to control as knowledge of the shelf life of many media may be lacking and reliable storage times will in any case vary according to conditions. However, some general guidelines are as follows:

Estimated shelf life A turnover period of about one month for most characterization media is probably a reasonable compromise between excessive storage times and the need to prepare batches too frequently. Batch sizes should be planned accordingly. Empty containers of tubes completely before adding tubes from a fresh batch. This prevents the possibility of the accumulation of a stagnant pool of tubes from previous batches and allows the date marking of containers rather than individual tubes. Record the date of preparation of each batch.

Temperature of storage Ensure storage conditions are cool (eg, not above a radiator). Labile media, for example ONPG broths, should be stored in the refrigerator. Evaporation must be prevented, either by the use of sealed plastic bags for tubed media or by using screw-capped containers (eg bijou bottles).

Humidity Excess humidity is a common cause of deterioration of dehydrated reagents. Containers removed from storage in the refrigerator must be allowed to warm up to room temperature before opening to prevent condensation of water vapour inside the container, with subsequent hydration of the reagent.

Long-term storage Media used for infrequently-performed tests may have long turnover periods. Batches of many commonly-used characterization media have been stored in the author's laboratory for periods of six months without deterioration in performance. However, it is important to check such rarely-used media with positive and negative controls each time they are used. In many cases simple visual inspection will reveal changes in colour or turbidity indicating deterioration. Some media and reagents which deteriorate with prolonged storage are KCN broth, V factor discs, ONPG broth, bacitracin discs, selenite broth, hydrogen peroxide, and some carbohydrate-containing media, eg arabinose, xylose and starch.

Purchase raw ingredients liable to deteriorate, such as peptones, in small quantities. Note the dates of receipt and opening on the bottle, and inspect the ingredients for evidence of deterioration, especially hardening caused by moisture, on each occasion of use.

Purity checks

Most gross contamination will become apparent after five days at room temperature and fresh batches of characterization media should be stored for this period before use. Incubate media at the normal temperature of incubation of the test before the batch is released for use. Sporadic contamination can only be detected by examining each tube for colour change or turbidity before inoculation and workers should be encouraged to do this. It is essential that the purity of the inoculum is checked. Inoculation of characterization tests with growth from selective media increases the risk of contamination and if the delay is permissible, it is safer first to sub-culture on to non-selective media. In all cases it is essential that the inoculum used for the tests is plated on a non-selective medium to check for purity. It is good practice to inoculate from a suspension of bacteria, the purity of which is easily checked by plating. Many environmental contaminants grow best at temperatures lower than 37°C and purity plates should be rechecked after being at room temperature following initial incubation. In medical microbiology laboratories, disc susceptibility tests are often performed in parallel with identification procedures and mixed populations may be easily seen in the inhibition zones around the discs.

Record keeping

The keeping of simple records provides assurance that quality control procedures have been carried out, aids the detection of faulty batches, encourages a methodical approach and helps to introduce the concept of accountability. Keep a media preparation card or book, recording the date of preparation of each batch or date of receipt of commercially-produced reagents, the size of batch and batch number, the pH of medium, the results of testing with control strains, comments and the operator's initials.

Use of control strains

Every characterization test should be controlled with both positive and negative control strains. Test media and reagents immediately after preparation and also after extended storage if the medium is infrequently used. Misreading test results due to lack of familiarity is an important cause of errors. The manufacturers of identification kits will often demonstrate how to read results obtained with their products and it is a good idea to take advantage of this service. A wide range of characterization tests is available and to control all of them would require the maintenance of a large number of strains. However, routine laboratories are unlikely to use all of the tests available and a smaller number can be selected to control only those tests in use. A set of four strains, available from the National Collection of Type Cultures (NCTC), controls the routinely-used tests listed in Table 3.1. Some other commonly-used tests and suggested controls, also available from the NCTC, are listed in Table 3.2. The NCTC catalogue is available from the Public Health Laboratory Service. Laboratories using other tests and dealing with specialized groups of organisms must select and maintain appropriate controls. Safety precautions must be observed in handling strains used for quality control purposes, since laboratory infections have been associated with the use of some strains, such as *Salmonella typhi*.[8]

Quality control in the microbiology laboratory

Table 3.1 NCTC control strains for commonly-used tests

Test	NCTC strain	
	positive	negative
Aesculin hydrolysis	11935	11934
Citrate utilization	7475	11934
Decarboxylases		
arginine	11936	7475
lysine	11935	7475
ornithine	11935	7475
Deoxyribonuclease	11935	11934
Gelatin liquefaction	11935	11936
Gluconate oxidation	11936	7475
Hydrogen sulphide (TSI)	11934	7475
Indole production	7475	11935
KCN tolerance	7475	11934
Malonate utilization	11936	7475
Methyl red	7475	11935
ONPG	11935	7475
PPA production	7475	11936
Selenite reduction	11936	11934
Urease	7475	11935

Test	NCTC strain	
	positive	negative
Voges Proskauer	11935	7475
Gas from glucose	11936	11935
Acid from sugars		
Adonitol	7475	11934
Arabinose	11936	11934
Dulcitol	11936	11934
Inositol	11935	11934
Lactose	11936	11934
Maltose	11936	7475
Mannitol	11936	11934
Raffinose	11936	11934
Rhamnose	11936	11934
Salicin	11935	11934
Sorbitol	11936	11934
Sucrose	11936	11934
Trehalose	11936	11934
Xylose	11936	11934

NCTC 11935 = *Serratia marcescens*
NCTC 11934 = *Edwardsiella tarda*
NCTC 7475 = *Proteus rettgeri*
NCTC 11936 = *Enterobacter cloacae*

Bacteriological characterization tests

Table 3.2 Further commonly-used tests and suggested controls

Test	Control organism	NCTC number	Expected result
Bacitracin sensitivity	*Streptococcus pyogenes*	8198	Sensitive
	Streptococcus viridans	10712	Resistant
Catalase	*Staphylococcus aureus*	853	Positive
	Streptococcus pyogenes	8198	Negative
Coagulase	*Staphylococcus aureus*	8532	Positive
	Staphylococcus epidermidis	4276	Negative
Deoxyribonuclease	*Staphylococcus aureus*	8532	Positive
	Staphylococcus epidermidis	4276	Negative
Haemolysis	*Streptococcus pyogenes*	8198	β-haemolysis
	Streptococcus viridans	10712	α-haemolysis
	Staphylococcus epidermidis	4276	no haemolysis
Oxidation/fermentation	*Pseudomonas aeruginosa*	10662	Oxidative
	Serratia marcescens	11935	Fermentative
	Acinetobacter lwoffii	5866	Alkaline or negative
Motility	*Serratia marcescens*	11935	Motile
	Acinetobacter lwoffii	5866	Non-motile
Nitrate reduction	*Serratia marcescens*	11935	Positive
	Acinetobacter lwoffii	5866	Negative
Optochin sensitivity	*Streptococcus pneumoniae*	10319	Sensitive
	Streptococcus viridans	10712	Resistant
Oxidase	*Pseudomonas aeruginosa*	10662	Positive
	Acinetobacter lwoffii	5688	Negative
Phosphatase	*Staphylococcus aureus*	8532	Positive
	Staphylococcus epidermidis	4276	Negative
Toxigenicity testing of			
C. diphtheriae	*Corynebacterium diphtheriae*	10648	Positive
	C. diphtheriae	3984	Weak positive
	C. diphtheriae	10356	Negative
X and V factor	*Haemophilus influenzae*	10479	Requires X and V
	Haemophilus parainfluenzae	10665	Requires V
	Haemophilus canis	8540	Requires X

References

1 Sneath PHA. Test reproducibility in relation to identification. *Int J Syst Bacteriol* 1974; **24**: 508–23.
2 Sneath PHA, Collins VG. A study in test reproducibility between laboratories: report of a pseudomonas working party. *Antonie Van Leeuwenhoek* 1974; **40**: 481–527.
3 Sneath PHA, Johnson R. The influence on numerical taxonomic similarities of errors in microbiological tests. *J Gen Microbiol* 1972; **72**: 377–92.
4 Lapage SP, Bascomb S, Willcox WR, Curtis MA. Identification of bacteria by computer: general aspects and perspectives. *J Gen Microbiol* 1973; **77**: 273–90.
5 Snell JJS, DeMello JV. Phua TJ. Errors in bacteriological techniques: results from the United Kingdom national external quality assessment scheme for microbiology. *Med Lab Sci* 1986; **43**: 344–55.
6 Hamilton WJ, Martin RJ. Survey of clinical microbiological techniques used in the United Kingdom. *Med Lab Technol* 1975; **32**: 307–12.
7 Cowan ST. Cowan and Steel's manual for the identification of medical bacteria. 2nd ed. Cambridge: Cambridge University Press, 1974.
8 Blaser MJ, Hickman FW, Farmer JJ, Brenner DJ, Balows A, Feldman RA. *Salmonella typhi*: the laboratory as a reservoir of infection. *J Infect Dis* 1980; **142**: 934–8.

4 Antibiotic susceptibility testing

DFJ Brown

Several types of susceptibility test are used in diagnostic microbiology laboratories.

Disc diffusion methods

Some form of disc diffusion test is used for routine susceptibility testing in most clinical microbiology laboratories. Comparative methods, including Stokes' modification,[1,2,3] are commonly used in the United Kingdom and allow more technical variation than the standardized methods[4,5,6] used in some other countries. Within individual laboratories, however, the technique is often standardized to a large extent.

Minimum inhibitory concentration (MIC) methods[7,8]

MIC methods are not commonly used for routine susceptibility testing, but may be appropriate when information more quantitative than that available from a diffusion test is desirable, or for tests on slow-growing organisms. Agar or broth dilution methods may be used. Most standardized techniques can be traced to those used in the World Health Organization International Collaborative Study of Susceptibility Testing,[9] but there is considerable variation among laboratories in the way they are performed. Again, such methods tend to be well standardized within individual laboratories.

Breakpoint methods[10]

These methods are used in some laboratories for routine susceptibility testing. Details of technique vary among laboratories but are well standardized within laboratories.

Quality control in the microbiology laboratory

Minimum bactericidal concentration (MBC) methods

The MBC may be useful in situations where knowledge of the bactericidal activity of an agent is important, eg in the treatment of endocarditis. The test is usually an extension of the broth dilution MIC method in that tubes showing no growth are sub-cultured and the number of colony-forming units compared with the number in the original inoculum.

Special methods

Tests may be based on detection of specific resistance mechanisms, eg production of β-lactamases or detection of chloramphenicol acetyl-transferase.

Different procedures are necessary to control the various methods used. When a method is standardized it is easier to use internal quality control tests to detect departures from normality which might give rise to errors. Moreover, as errors become rarer they are less likely to be detected by external quality control. The degree of control depends on resources, perception of the need for quality control and interest. It is important that the likely sources of error in a method are known so that if control procedures indicate a problem, the source of any error can be readily detected and rectified.

Control procedures

Control strains

Characteristics of control strains   Control procedures for all methods should involve the use of control strains. Strains sensitive to antimicrobial agents are commonly used but resistant strains are necessary in tests for resistance mediated by inactivating enzymes and for breakpoint methods. The susceptibility of control strains must be stable during long-term storage. The strains should grow well on the media used for susceptibility testing and should give clear zone edges in diffusion tests or sharp endpoints in MIC tests. The control strains should also give reproducible results when the test is performing correctly.

Storage of control strains[11]   Control cultures should be obtained from a reliable source, preferably an established culture collection. The susceptibility of the strains should be stabilized by careful preservation (see Chapter 7). Strains that have acquired resistance by mutation or by acquisition of plasmids may be particularly unstable.

Comparative disc diffusion methods

Control strains   In the comparative methods, the control strains serve two functions. Firstly they are the basis of interpretation, in that the zone sizes of test strains are compared with the zone sizes of control strains set up at the same time. Secondly they are the means by which variation in the test is controlled, in that any variations in the medium, disc content or incubation conditions are assumed to affect the test and control similarly and therefore cancel out. Hence, if the comparative methods are to be interpreted correctly, control organisms must be used. In the original descriptions of the comparative methods, the control organism for isolates from systemic infections was *Staphylococcus aureus* NCTC 6571, and the control organism for isolates from urinary tract infections was *Escherichia coli* NCTC 10418. The exception was for strains of *Pseudomonas* spp which were compared with *Pseudomonas aeruginosa* NCTC 10662. Whether the *Escherichia coli* and *Staphylococcus aureus* controls are entirely suitable has been questioned since they are particularly sensitive to some agents. Even so, alternatives have not yet been proposed.

More recently, there has been a trend towards the use of control strains more typical of the isolate rather than related to the source of the isolate. There has not, however, been any formal validation of this approach. Several control strains have been recommended:[1,2,3]

- *Escherichia coli* NCTC 10418 for tests on coliform organisms
- *Pseudomonas aeruginosa* NCTC 10662 for tests on *Pseudomonas* spp
- *Haemophilus influenzae* NCTC 11931 for tests on *Haemophilus* spp[12]
- *Neisseria gonorrhoeae* (unspecified sensitive strain) for tests on *Neisseria gonorrhoeae*[13]
- *Staphylococcus aureus* NCTC 6571 for tests on other organisms that grow aerobically
- *Clostridium perfringens* NCTC 11229 for tests on *Clostridium* spp[14]
- *Bacteroides fragilis* NCTC 9343 for tests on other rapidly growing anaerobic organisms[15]. This strain, however, is penicillin resistant and therefore cannot be used for tests with penicillin.
- *Escherichia coli* NCTC 11560, a strain which produces β-lactamase, may be used for quality control of the amount of β-lactamase inhibitor in discs containing a combination of a β-lactam agent and a β-lactamase inhibitor
- *Streptococcus pneumoniae* NCTC 12140 is moderately resistant to benzyl penicillin (MIC 0.25mg/l) and may be used for quality

control of tests for decreased susceptibility to penicillins in pneumococci.[16]

Required procedures for comparative methods The Stokes method requires every test to have a control strain on the same plate as the test. Other comparative methods use a separate plate for the control, which should also be set up daily.

As the method of interpretation is assumed to control variation in the test, no action generally need be taken when control zones vary. Although no limits for control zone sizes have been published, it is expected that some action would be taken if control zone sizes were wildly aberrant. The reproducibility of sizes of zones of inhibition with control strains indicates how the method is performing. A cursory examination of control zones is clearly a minimum need and will indicate gross abnormalities. Some common problems are described in the section on sources of errors.

Other disc diffusion methods

Control strains Several standardized methods have followed the lead of the United States National Committee for Clinical Laboratory Standards (NCCLS) with regard to the specific control strains used. A variety of strains are recommended for tests on organisms that grow aerobically. NCCLS do not advise testing anaerobic organisms by diffusion methods. Control strains recommended are:

- *Escherichia coli* ATCC 25922 (NCTC 12241) for regular use as a control
- *Staphylococcus aureus* ATCC 25923 for regular use as a control
- *Pseudomonas aeruginosa* ATCC 27853 for regular use as a control
- *Haemophilus influenzae* ATCC 49247 for regular use as a control
- *Neisseria gonorrhoeae* ATCC 49226 for regular use as a control
- *Escherichia coli* ATCC 35218 (NCTC 11954), a strain which produces β-lactamase, may be used for quality control of the amount of β-lactamase inhibitor in discs containing a combination of a β-lactam agent and a β-lactamase inhibitor
- *Enterococcus faecalis* ATCC 29212 or ATCC 33186 may be used to test for sulphonamide/trimethoprim antagonists.

Required procedures for the NCCLS method The NCCLS requires that until it is established that the test is performing well, controls should be set up daily. Once published performance criteria are met[5], the

Antibiotic susceptibility testing

frequency of testing controls may be reduced to once a week. In addition, controls should be tested whenever there is a change, such as a new batch of medium or discs. Any sign of reduced accuracy, indicated by control zone diameters outside the ranges published by NCCLS,[5] should result in a return to daily testing of controls until the problem is resolved.

The NCCLS give statistics for acceptability of control zone sizes as follows.

1 A single test should have a zone diameter within the given range. If more than one in twenty results is outside the given range (equivalent to 95% confidence limits), corrective action is considered necessary.

2 Control zone sizes should never be more than four standard deviations above or below the midpoint of the acceptable range.

3 The mean control zone size should be close to the midpoint of the acceptable range.

4 The range of a series of five consecutive control zone sizes should not exceed stated limits.

Required procedures for other disc diffusion methods Most other standardized methods require daily controls. Tables of expected control zone diameters have been published for some methods although acceptable limits of control zone sizes are not generally available.

Additional control measures for disc diffusion methods

1 With all methods, recording sizes of control zones daily on a chart makes departures from normality easier to see. If control zone limits are available they can be marked on the chart. A gradual reduction in zone size with labile agents such as penicillins and cephalosporins suggests deterioration of the agent in the discs due to inadequate storage or handling of the discs. Generally, larger or smaller zones may indicate too light or too heavy an inoculum respectively. Larger zones may also indicate that the depth of medium in the plates is too shallow. Larger zones with aminoglycosides at the same time as smaller zones with tetracycline suggests that the pH of the medium is too high; the reverse may be seen if the pH is too low. Sudden changes in zone size may indicate problems with a new batch of medium or discs, or observer variation in the reading of zone edges.

2 As with the NCCLS method, greater control over test performance

may be achieved by establishing limits for sizes of control zones. Such limits have not been established for the 'comparative' methods or for most other standardized methods. If external quality control suggests that the method used is reliable, acceptable ranges of sizes of control zones may be established for routine use by calculating 95% confidence limits. This is achieved by calculating the standard deviation of a series (as large as possible, but a minimum of 30) of sizes of control zones. The 95% confidence limits, and hence the acceptable range for nineteen out of twenty sizes of control zones, is the mean zone size plus or minus two standard deviations.

3 Deterioration of a β-lactamase inhibitor in discs containing a combination of a β-lactam agent and a β-lactamase inhibitor may be recognized by the reduction in size of zones with β-lactamase producing strains *Escherichia coli* ATCC 35218 (NCTC 119540) or *Escherichia coli* NCTC 11560. The β-lactam component in such discs may be monitored by testing with *Escherichia coli* NCTC 10418 or *Escherichia coli* ATCC 25922 (NCTC 12241).

4 New batches of medium should be tested for unacceptable amounts of sulphonamide/trimethoprim antagonists (principally thymidine and thymine) with *Enterococcus faecalis* strains ATCC 29212 or ATCC 33186. Unclear zones of inhibition around trimethoprim or cotrimoxazole discs with this organism indicate that the medium is unsatisfactory.

5 Methods of testing for methicillin/oxacillin resistance in staphylococci should include a sensitive control strain. A resistant strain, *Staphylococcus aureus* MQCL 113, may also be used to check that test conditions are favourable for the detection of resistance. Colonies of the resistant control should grow up to the disc.

Minimum inhibitory concentration methods

Control strains   The MIC of the control strain should fall within the range of concentrations tested. In general the strains used for control of diffusion tests are used as controls in MIC tests. The NCCLS have recommended *Staphylococcus aureus* ATCC 29213 as an alternative to *Staphylococcus aureus* ATCC 25923 because the latter is very susceptible to many agents.[8] It may be useful to include control strains that represent species being tested, particularly when media or incubation conditions need to be altered to permit growth, but a wide range of control strains has not been recommended.

Antibiotic susceptibility testing

For tests on anaerobic organisms, NCCLS recommend that at least one of the following control strains is used:[17]

- *Bacteroides fragilis* ATCC 25285 (NCTC 9343)
- *Bacteroides thetaiotaomicron* ATCC 29741
- *Clostridium perfringens* ATCC 13124 (NCTC 8237)
- *Eubacterium lentum* ATCC 43055.

Control procedures Control strains should be included with each batch of tests and control values recorded.

1 The MICs of the control strain should be known. The NCCLS have published ranges of MICs for standard strains tested by an agar dilution method on Mueller–Hinton agar.[8] Modal MICs for control strains commonly used in the UK have been published.[18] Table 4.1 gives MICs for control organisms commonly used and tested on Iso-Sensitest agar (Oxoid). Details of techniques used in various laboratories may be different and conditions used for various tests within the same laboratory may be different, eg the medium may be supplemented or the atmosphere for incubation may be altered for fastidious organisms. MICs may vary slightly depending on details of the method used. Unless the NCCLS method is followed exactly, control values should be established for the method in use. Results of control MICs should be accumulated from repeated tests with the control strains over a period of time. The modal MIC values should thereby be established and should not differ markedly from published values obtained with other methods (where such data are available). The MIC should not vary by more than one doubling dilution value above or below the mode. If the MIC falls outside this range, sources of error should be sought.

2 In each test a control without antibiotic should be included so as to ensure that the test strain grows adequately.

3 Purity of inoculum should be tested by streaking samples of the inoculum on suitable medium so as to obtain isolated colonies. In broth dilution tests this is essential since mixed cultures are likely to go unrecognized.

Breakpoint tests

Control strains Selection of control strains for the breakpoint method is difficult. The one or two antibiotic concentrations tested often differ considerably from the MICs of control strains used in disc diffusion and MIC methods. Hence the control strains used in the latter methods are of little value in breakpoint tests. Control strains suitable

Quality control in the microbiology laboratory

Table 4.1   MICs for control strains

Antimicrobial agent	E coli NCTC 10418	Ps aeruginosa NCTC 10662	S aureus NCTC 6571
Amikacin	1	2	1
Amoxicillin	2	–	0.06
Ampicillin	2	–	0.06
Azlocillin	8	4	0.25
Aztreonam	0.03	4	–
Benzyl penicillin	–	–	0.03
Cefalexin	8	–	1
Cefotaxime	0.03	8	1
Cefoxitin	1	–	1
Ceftizoxime	0.015	–	2
Cefradine	8	–	2
Cefsulodin	–	2	–
Ceftazidime	0.125	1	4
Cefuroxime	2	–	0.5
Chloramphenicol	2	–	4
Ciprofloxacin	0.015	0.125	0.125
Clindamycin	–	–	0.06
Erythromycin	–	–	0.125
Fusidic acid	–	–	0.06
Gentamicin	0.25	1	0.125
Kanamycin	2	–	0.5
Methicillin	–	–	1
Mezlocillin	2	8	0.5
Netilmicin	0.5	1	0.25
Nalidixic acid	1	–	–
Nitrofurantoin	4	–	–
Oxacillin	–	–	0.25
Piperacillin	1	2	0.5
Rifampicin	–	–	0.008
Sulfamethoxazole	8	–	8
Tetracycline	1	–	0.125
Ticarcillin	1	16	0.5
Tobramycin	0.25	0.5	0.125
Trimethoprim	0.125	–	0.5
Trimethoprim/ sulfamethoxazole (1:20)	0.01/2	–	0.05/1
Vancomycin	–	–	0.5

Modal values from determinations on Iso–Sensitest agar (Oxoid) in the author's laboratory.

Antibiotic susceptibility testing

for breakpoint tests should have MICs just above and just below each breakpoint, but not so close to the breakpoint that the normal variation in the test leads to day-to-day variation in results. In order to achieve this, it is likely that control strains will differ for the various agents. Consequently a range of control strains will be necessary if several agents are being tested. In practice, the required number of controls might severely restrict space on the plates for test organisms.

A set of control strains for use with particular breakpoints, relevant to isolates from the urinary tract tested by a defined technique, is being evaluated.[19] Strains with wider application are not yet available.

Control procedures
1   Control strains should be included with each batch of tests, but the limitations of the range of controls used should be recognized. Any variation in results of tests with control strains may indicate an error in the test and should be investigated.
2   Batch controls are a more practical approach to the adequate control of breakpoint tests. Plates are commonly poured in quantities large enough for a week's tests. An extra set of plates should be poured with each batch and used as a representative in quality control tests with a range of organisms to cover all breakpoints as described above. Any breakpoint giving an abnormal result with any control strain may indicate an error in the test and should be investigated.
3   The last set of plates in a batch can be used in parallel with the first set in the next batch. Discrepancies should be investigated and may indicate loss of potency during storage of the old batch or a production error with the new batch.

Minimum bactericidal concentrations

Control strains   The control strains used are the same as in the MIC methods.

Control procedures
1   The purpose of the control organism in MBC tests is simply to confirm that the antimicrobial concentrations are correct, as with MIC methods. Hence sub-culture of the control strain from tubes showing no growth is not necessary.
2   In each test a control without antimicrobial agent is included. Immediately after inoculation of this control tube, the tube is

Quality control in the microbiology laboratory

quantitatively sub-cultured on solid medium. The density of the inoculum can then be calculated from the number of colonies growing after incubation of the plate. The number of surviving organisms in tubes with antimicrobial agent can then be compared with the number of organisms in the original inoculum to establish the MBC (>99.9% kill). The purity of the inoculum can also be seen from the sub-cultures.

Special methods

Control strains   Negative and positive control strains are necessary. A suitable negative control would be a sensitive strain of the same genus as that being tested; an ideal positive control would be a resistant strain of the same genus but one which gives a weak positive result. For some tests, defined strains suitable for use as positive controls have not yet been deposited in culture collections but are available from the Quality Assurance Laboratory, Colindale (strains with MQCL numbers). Control strains recommended are:

- *Haemophilus influenzae* strains NCTC 11315 and NCTC 11931 are, respectively, positive and negative controls for β-lactamase tests on *Haemophilus* spp
- *Neisseria gonorrhoeae* strains NCTC 11148 and NCTC 8375 are, respectively, positive and negative controls for β-lactamase tests on gonococci
- *Haemophilus influenzae* strains MQCL 1136 and NCTC 11931 are, respectively, positive and negative controls for chloramphenicol acetylase tests on *Haemophilus* spp
- *Streptococcus pneumoniae* strains MQCL 1441 and NCTC 12140 are, respectively, positive and negative controls for chloramphenicol acetylase tests on pneumococci.

Control procedures   A positive and a negative control strain should be included in each batch of tests and should give the appropriate results.

Atypical results

Results of susceptibility tests with certain combinations of organism and antimicrobial agent are predictable because some species are inherently resistant to some agents or because acquired resistance is rare (although this may vary from one area to another and is likely to change with time). Unusual patterns of results may therefore indicate

Antibiotic susceptibility testing

error in the test, such as misidentification or the testing of a mixed culture. All possibilities should be investigated and if necessary the test repeated before the result is reported. This type of control may be particularly important with breakpoint tests, in which adequate daily control in all tests may be difficult to achieve with control organisms. If reporting is computerized, this type of control may be written into the programs.

Some examples of suspect reports are given below. Some of the combinations quoted are clinically inappropriate but may be tested through the use of standard sets of antimicrobial agents.

1   Apparent resistance in a species with which resistance has not previously been seen (eg penicillin resistance in Group A streptococci).
2   Streptococci apparently sensitive to aminoglycosides.
3   *Enterococcus faecalis* apparently sensitive to sulphonamides.
4   Pneumococci apparently resistant to penicillin or chloramphenicol.
5   Staphylococci apparently resistant to methicillin and sensitive to penicillin or cephalosporins.
6   Staphylococci apparently resistant to penicillin and sensitive to ampicillin.
7   Staphylococci resistant to lincomycin or clindamycin but sensitive to erythromycin.
8   Gram-positive organisms (excluding *Leuconostoc* spp, *Pediococcus* spp and lactobacilli) apparently resistant to vancomycin.
9   *Enterobacteriaceae* apparently resistant to augmentin but susceptible to amoxycillin.
10 *Enterobacteriaceae* apparently resistant to second and third generation cephalosporins but sensitive to first generation cephalosporins.
11 *Serratia* spp, *Yersinia enterocolitica* and *Citrobacter diversus* apparently sensitive to ampicillin.
12 *Serratia* spp, *Proteus* spp and *Providencia stuartii* apparently sensitive to colistin.
13 *Proteus* spp apparently sensitive to nitrofurantoin.
14 *Escherichia coli* apparently resistant to third generation cephalosporins.
15 *Morganella* spp apparently sensitive to augmentin and first generation cephalosporins.
16 *Pseudomonas aeruginosa* apparently sensitive to first generation

cephalosporins, nalidixic acid, trimethoprim and sulphon-amides.

17 *Campylobacter jejuni* apparently sensitive to vancomycin, trimethoprim and first generation cephalosporins.

18 *Haemophilus influenzae* apparently resistant to trimethoprim (check the medium) or chloramphenicol.

19 *Haemophilus influenzae* and *Neisseria gonorrhoeae* apparently resistant to penicillin when β-lactamase is not produced.

20 *Neisseria gonorrhoeae* and *Branhamella catarrhalis* apparently sensitive to trimethoprim.

Sources of errors

If the control procedure indicates that there may be an error in the test the problem should be investigated. Knowledge of the potential sources of error is therefore essential. Common sources of error are discussed below.

In diffusion tests by disc methods

Error in measuring sizes of control zones or transcription errors in recording sizes of control zones   If the zone edge is not sharp, defining the edge of the zone is liable to variation. The problem may be compounded by variable illumination. It may be rectified by re-examining control plates, standardizing illumination, and ensuring that all observers define zone edges according to established practice. If errors in measurement and transcription are frequent they probably arise in test strains also and may indicate a less than acceptable standard of performance of the test in general.

Contamination or mutational changes in the control strain   Take a new vial of the control organism from stock or obtain a fresh culture from an appropriate culture collection.

Problems with the medium due to improper preparation   Ensure that the medium is prepared as recommended by the manufacturer.

Problems with a new lot of medium due to variation among batches   Obtain an acceptable batch from the manufacturer. When possible, the purchase of large amounts of a batch found to give acceptable results should be considered. A medium other than the sensitivity testing medium may have been inadvertently used.

Antibiotic susceptibility testing

Problems with the antimicrobial content of discs Manufacturers are aware of their responsibility to impregnate and label discs accurately. Discs of unacceptable quality due to excessively high or low content, or discs containing the wrong agent, are now not often encountered. Labile agents in discs may deteriorate as a result of mishandling or unsuitable storage in the laboratory. This may be a particular problem in smaller laboratories where the turnover of discs may be slow. Store discs in airtight containers over an indicating desiccant. Store stocks at −20°C if possible, otherwise at <8°C. Working supplies should be stored at <8°C. Exposure of discs to moisture by opening vials of discs immediately after removal from refrigeration should be avoided. Vials should always be allowed to warm to room temperature before opening them. Metronidazole is particularly sensitive to light and should therefore be stored in the dark.

Inoculum too heavy or too light Check and if necessary adjust the method of preparing the inoculum. If a turbidity standard is used, the preparation and storage of this should be investigated.

Plates left on the bench in a warm laboratory before discs are applied This may result in growth of the organism and produce the same effect as increasing the inoculum. Ensure that discs are applied soon after plates are inoculated.

Delay in incubating plates after discs have been applied This may lead to enlarged zone sizes. Ensure that plates are incubated soon after discs are applied.

In MIC methods

Use of incorrect forms of agents Some pharmaceutical preparations include substances such as buffers which may lead to errors in weighing. Pure preparations should be obtained from the manufacturer or commercial sources. The agent should be supplied with a statement of the potency, expiry date, storage conditions and solubility.

Arithmetical errors in the preparation of dilution series Write clear protocols and ensure that staff understand them.

Failure to allow for the potency of preparations Pure preparations of agents from the manufacturer or from a chemical company should be

Quality control in the microbiology laboratory

supplied with stated potency. Allowance must be made for the potency of the preparation when weighing the drug.

Errors in weighing powders   When possible, at least 100mg powder should be weighed on an analytical balance, the accuracy of which is regularly checked.

Inappropriate solvents   Most agents may be dissolved and diluted in water. Some require other solvents.[8]

Incorrect storage of powders or stock solutions   Store powders as recommended by the manufacturer. Most agents are stable if stored at $-20°C$ or below in a desiccator. Containers should be opened only when the contents are at ambient temperature. Containers should be open for as short a period as possible to avoid absorption of moisture, which may lead to deterioration of the agent and to errors in weighing. Store stock solutions at $-20°C$ or below (preferably $-60°C$) in small amounts. Avoid repeated freezing and thawing of solutions.

Errors in pipetting solutions or failure of diluting systems   Check the accuracy of any automated dilution system periodically.

Inactivation of agents by the use of hot molten agar   Agar should be cooled to 50°C before being mixed with antimicrobial agents. Plates should be poured as soon as possible after addition of agents to molten agar.

Incorrect storage of plates or broth containing agents   Agar plates containing antimicrobial agents may be stored in sealed bags at 4°C for one week. Broth containing antimicrobial agents may be frozen at $-60°C$ in microtitre trays for one month, or in sealed tubes at 4°C for one week.

Errors due to problems with the medium   If the medium is supplemented in order to allow the growth of fastidious organisms, the inclusion of controls should ensure that MICs are not adversely affected. High pH increases the activity of aminoglycosides and macrolides but decreases the activity of tetracyclines; low pH has the opposite effect.

Incorrect inoculum   Increasing the inoculum usually raises the MIC, whereas decreasing the inoculum has the reverse effect.

Spreading or running of inoculum spots on agar dilution plates This may arise with undried plates or when plates have been moved or inverted before spots have dried.

Failure of mechanical inoculators This may result in tubes or spots on a plate being missed, or in contamination due to splashing.

Gross contamination of inoculum wells This may be due to splashing when filling wells. To avoid splashing, pipettes or micropipettes should not be fully expelled.

Incorrect temperature or time of incubation The temperature and time of incubation should be as recommended for the method used.

Errors in reading endpoints or transcription errors Staff should be instructed in reading of endpoints. Single colonies or a haze of growth are generally disregarded.

Contamination or mutational changes in the control strain Take a new vial of the control organism from stock or obtain a fresh culture from an appropriate culture collection.

In breakpoint methods

Errors common in breakpoint methods are similar to those described above for MIC tests. Some laboratories use pre-weighed antibiotics in the form of Adatabs® (Mast). Major problems with the reliability of these products have not been recorded. In view of the lack of discriminating control of this method in many laboratories, however, only larger errors are likely to be noticed.

References

1　Stokes EJ, Waterworth PM. Antibiotic sensitivity tests by diffusion methods. Association of Clinical Pathologists Broadsheet 55, 1972.
2　Waterworth PM. Laboratory control. In: Garrod LP, Lambert HP, O'Grady F, eds: Antibiotic and chemotherapy. Edinburgh: Churchill Livingstone, 1981; 464–504.
3　Stokes EJ, Ridgeway GL. Clinical bacteriology, 6th ed. London: Arnold, 1987; 204.
4　World Health Organization. Requirements for antimicrobic susceptibility tests I. Agar diffusion tests using antimicrobic susceptibility discs. Technical Report Series, 673, Annex 5. Geneva: World Health Organization, 1982; 144–78.

Quality control in the microbiology laboratory

5 National Committee for Clinical Laboratory Standards. Performance standards for antimicrobial disk susceptibility tests, 4th ed. Approved standard M2–A4. Villanova, PA: NCCLS, 1990.

6 European Committee for Clinical Laboratory Standards. Guidelines for antimicrobial susceptibility testing by diffusion methods. ECCLS Document 1: Lund, 1988.

7 Waterworth PM. Quantitative methods for bacterial sensitivity testing. In: Reeves DS, Phillips I, Williams JD, Wise R, eds: Laboratory methods in antimicrobial chemotherapy. Edinburgh: Churchill Livingstone, 1978; 31–40.

8 National Committee for Clinical Laboratory Standards. Methods for dilution antimicrobial susceptibility tests for bacteria that grow aerobically, 2nd ed. Approved standard M7-A2. Villanova, PA: NCCLS, 1990.

9 Ericsson H, Sherris JC. Antibiotic sensitivity testing. Report of an international collaborative study. *Acta Pathol Microbiol Scand* [B]1971; supplement no 217.

10 Waterworth PM. Sensitivity testing by the breakpoint method. *J Antimicrob Chemother* 1983; **7**: 117–26.

11 Kirsop BE, Snell JJS, eds. Maintenance of microorganisms. A manual of laboratory methods. London: Academic Press, 1984.

12 Snell JJS, Brown DFJ, Phua R. Antimicrobial susceptibility testing of *Haemophilus influenzae*: a trial organised as part of the United Kingdom national external quality assessment scheme for microbiology. *J Clin Pathol* 1986; **39**: 1006–12.

13 Jephcott AE, Eggleston SI. *Neisseria gonorrhoeae*. In: Collins CH, Grange JM, eds: Isolation and identification of microorganisms of medical and veterinary importance. London: Academic Press, 1985; 150–3.

14 Milne SE, Stokes EJ, Waterworth PM. Incomplete anaerobiosis as a cause of metronidazole resistance. *J Clin Pathol* 1978; **7**: 933–5.

15 Phillips I, Warren C. Anaerobic bacteria. In: Reeves DS, Phillips I, Williams JD, Wise R, eds: Laboratory methods in antimicrobial chemotherapy. Edinburgh: Churchill Livingstone, 1978; 95–8.

16 Snell JJS, George RC, Perry SF, Erdman YJ. Antimicrobial susceptibility testing of *Streptococcus pneumoniae*: quality assessment results. *J Clin Pathol* 1988; **41**: 384–7.

17 National Committee for Clinical Laboratory Standards. Methods for antimicrobial susceptibility testing of anaerobic bacteria, 2nd ed. Tentative standard M11-T2. Villanova, PA: NCCLS, 1989.

18 Phillips I, Williams D. Antimicrobial susceptibility testing. In: Reeves DS, Phillips I, Williams JD, Wise R, eds: Laboratory methods in antimicrobial chemotherapy. Edinburgh: Churchill Livingstone, 1978; 3–7.

19 George RC, Warner M. BSMT study of breakpoint susceptibility testing of bacteria isolated from lower urinary tract infection. *British Society for Multipoint Technology Newsletter* no 6, 1987; 20.

5 Antibiotic assays

LO White and DS Reeves

Background

Some antibiotics have a narrow margin of safety between toxic and therapeutic serum concentrations. It is mandatory to assay these antibiotics as an aid to the correct management of patients. The most frequently requested assays are certain aminoglycosides (amikacin, gentamicin, netilmicin, tobramycin), vancomycin, chloramphenicol and the antifungal agent, flucytosine. There are, however, others which require monitoring in some patients but which are encountered less frequently. These include streptomycin, co-trimoxazole (trimethoprim and sulphamethoxazole), and benzyl penicillin. Twenty years ago almost all clinical antibiotic measurements were performed using microbiological assay but these were generally too slow for therapeutic monitoring purposes. Very rapid and reliable commercial immunoassay kits are now available for most commonly assayed antibiotics and almost all clinical laboratories use them. For those drugs where kits are not available many laboratories use high performance liquid chromatographic (HPLC) methods which can also produce a rapid result.

No matter what antibiotic is being assayed, quality control procedures are required to assure, with a high level of confidence, the validity of the results and thus the correctness of any related clinical advice. Quality control procedures are only a part of overall quality assurance. Aspects of quality assurance outside the scope of this chapter include exhaustive assay validation (which particularly applies to in-house methods), service and maintenance of equipment,

Quality control in the microbiology laboratory

proper record keeping, avoiding unacceptable delays between the receipt of the sample and the generation of the report, and ensuring that the service is not abused.

It is usually thought that to be clinically useful an assay technique should be able to reliably produce results within 25% of a 'true' value. Therefore a result of 2mg/l gentamicin can be taken by the clinician to have a true value somewhere between 1.5 and 2.5mg/l. Three factors can contribute to making an actual result the true result. The first is accuracy which is also referred to as bias, the second is precision and the third is specificity. Quality (which is a degree of excellence) is subjectively determined from knowledge of accuracy, precision and specificity.

Definitions

Accuracy is a measure of the closeness of a result to a true value when the same sample is assayed repeatedly. If a particular result is always lower than a true value the assay is said to have a negative bias. If the results are always higher than a true value the assay has a positive bias.

Precision is a measure of the closeness of agreement between results when the same sample is assayed several times. Precision is usually determined by assaying at least six replicates and determining the mean result and the standard deviation. Degrees of precision at different concentrations are compared by calculating the coefficient of variation (%), which is the standard deviation expressed as a percentage of the mean.

Specificity is a measure of a system's ability to measure the correct analyte and not to measure something else mistakenly. Assay specificity is measured by its ability to give accurate results in the presence of other analytes. Microbiological assays may show poor specificity if an indicator organism is inhibited by antibiotics in the sample other than the one being assayed. Immunoassays are usually highly specific because they rely on antigen-antibody interactions which are themselves highly specific. The degree of specificity of an assay should be determined at validation.

The following definitions are based on the 1990 definitions of the European Committee for Standardization with some enhancements relevant to antibiotic assays.[1]

Calibrator. Material used for calibration of a measurement procedure and to calculate the result of an analysis. Most antibiotic assays will require more than one calibrator (up to six) and each calibrator comprises a known amount of antibiotic dissolved in an appropriate matrix. (Note: we have previously called calibrators, working standards.)

Calibration The set of operations which establish, under specified conditions, the relationship between values indicated by a measuring instrument or system and the corresponding values of a quantity realised by a reference standard or standards (of known concentration in the case of antibiotic assays).

Internal quality control A procedure which utilises the results of only one laboratory for quality control purposes. Note that 1) an internal quality control sample may contain more than one analyte, 2) the acceptable result for some internal quality control samples of commercial origin may be method specific.

External quality assessment (EQA) A procedure which utilises, for quality control purposes, the results of several laboratories which assay the same specimen(s). Note that an EQA sample may contain more than one analyte. EQA samples may have target values based on actual weighed-in amounts or on consensus mean values (trimmed or untrimmed).

Control materials Any material used for internal quality control or external quality assessment.

Quality control of assays has three facets which will be discussed in turn. These are: assay protocols and calibration; internal quality control; and external quality assessment.

Assay protocols and calibration

Every assay should be performed according to strict protocols laid down in a laboratory manual or the laboratory standard operating procedures. The degree of specificity of an assay will have been determined at validation. Microbiological assays suffer specificity problems because multiple antibiotic therapy is common. These prob-

Quality control in the microbiology laboratory

lems are overcome in various ways, for example by the use of a multi-resistant indicator organism such as *Klebsiella edwardsii* NCTC 10896 for aminoglycoside assays. Some HPLC assays suffer specificity problems caused by unidentified substances co-eluting with the drug being measured. A simple way to monitor the specificity of HPLC results is to use in series two detectors set at different wavelengths. If the peak seen from the sample is the drug being assayed then the results obtained from the two detectors will agree with one another. If it is actually an interfering peak the results from the two detectors will probably disagree. This is because different compounds do not usually have the same absorption spectra. Immunoassays are usually highly specific. It should be noted, however, that the TDX assay for vancomycin cross-reacts with inactive vancomycin breakdown products, and in other immunoassays gentamicin cross-reacts with netilmicin and tobramycin with amikacin.

Whatever assay is being performed, all reagents should be prepared exactly as defined in the procedures. This applies to kit reagents which require reconstitution. The exact volumes and exact reconstituting fluids as laid down by the kit manufacturer should be used. Calibrators which require reconstitution should be prepared likewise. Once reconstituted, any 'ripening' period for reagents laid down by the manufacturer should be observed.

All reagents and calibrators should be stored according to laboratory or the manufacturer's instructions. For example, Abbott TDX vancomycin calibrators are shipped frozen, and should be thawed and then stored in a refrigerator for a maximum of one month. Abbott state that you should *not* refreeze the calibrators once thawed; standard curves generated from refrozen calibrators stay within specification for a shorter period of time. Since the cost of generating a standard curve is more than the cost of a set of calibrators, refreezing in not cost-effective. Expiry dates must be observed.

If 'in-house' methods are being used then two factors are essential for good accuracy in addition to good specificity and precision: correctly prepared calibrators and appropriate curve fitting procedures. It should also be noted that the second point also applies to commercial kits used in 'non-standard' procedures, for example EMIT assays performed on a variety of different autoanalysers.

Calibrators prepared 'in-house' should be prepared using a powder or solution supplied for the purpose by the relevant pharmaceutical company. The powder or solution should have a declared potency

Antibiotic assays

and preferably be an 'analytical reference material'. If it has an expiry date and specified storage instructions, these should be adhered to. Care must be taken when weighing out powders. Containers should be allowed to reach room temperature before they are opened to eliminate the possibility of water condensation on the dry powder. Some antibiotic powders may absorb water from the atmosphere. For example, gentamicin powder may have an anhydrous potency value. If this is so, the powder is heated at 110°C for 60 minutes and cooled in a dessicator before weighing. A suitable balance must be used; it should be capable of weighing 50 milligrams of powder accurately and quickly to three significant figures. Solutions should always be made up in Grade A volumetric glassware. Standard solutions for calibration purposes must be prepared in an appropriate matrix. For most microbiological assays this should be pooled human serum for assaying drugs in serum, although for some assays an animal substitute may be acceptable but this will have been determined during assay validation. Microbiological assays usually require five standards with concentration increasing in a two-fold manner to cover a therapeutic range. These standards must not be prepared by serial two-fold dilution since any dilution error will be magnified with each dilution. HPLC procedures usually demand fewer standards and the use of a matrix resembling the sample may not be obligatory. Some drugs bind to plastic or glass and attempts should be made to minimise this by, for example, not dispensing relatively small volumes in large containers. Ideally, all containers should be at least half full. Once prepared, standards may be used over a long time. Many drugs are stable for several months at -20°C (for example, most aminoglycosides, quinolones, vancomycin, teicoplanin, flucytosine, chloramphenicol). Others (for example, streptomycin and many beta-lactams) are relatively unstable. A laboratory doing infrequent assays for, say, benzyl penicillin would be advised to make up fresh standards each time the assay is run rather than storing them for any length of time.

Curve fitting is an aspect of quality control which is often overlooked. For many microbiological assays the square of the zone diameter is proportional to the logarithm of the concentration over a wide range of concentrations. Standard curves are, however, usually constructed by plotting logarithm of concentration against zone diameter. These curves may often look relatively linear but are in fact slightly curved. For clinical assays where the standards might well cover a 16-fold range of concentrations the curvature is usually quite obvious. Forcing a straight line through the points is inappropriate and will lead to assay bias at various points on the curve. A mathematical curve fit procedure will give a more accurate standard curve and various curve-fitting procedures can be found in the literature. At Southmead we have, for many years, recommended Perkins'[2] or Bennett's method and have programmed a number of micro-computers to do such calculations.

Curve fitting immunoassays often require complex mathematics. Thankfully, commercial immunoassays run on the appropriate equipment do not usually present curve-fit problems. Instruments such as the Abbott TDX or the Syva Clinical Processor have built-in computers which curve fit using appropriate equations. With assays run under non-standard conditions or on other analysers problems may be encountered. A good example is the EMIT gentamicin assay run on the Technicon RA 500. EMIT assays run on the RA 500 are calibrated normally and are curve-fitted using option 1A, a log/logit cubic transformation. This equation makes a relatively poor job of fitting EMIT curves and the result is a positive bias in the middle of the curve and a negative bias at the top end (Table 5.1). This bias can be >10% at 6mg/l and although not particularly significant for clinical assays, it rules out EMIT on the RA 500 as an acceptable procedure for pharmacokinetic studies.

Curve fitting chromatographic assays is not usually a problem

Table 5.1 A comparison of the Technicon RA 500 and Syva Clinical Processor curve fitting abilities

Concentration (mg/l)	Fit from RA 500 (mg/l)	Fit from Clinical Processor (mg/l)
0.0	0.0	0.0
0.6	0.6	0.6
2.0	2.02	2.0
4.0	3.79	4.0
6.0	6.69	6.1
10.0	9.38	9.9

The standards were run on the RA 500 and the fitted curve yielded the results as shown in column 2. The raw absorbance data was then entered manually into the clinical processor and used to construct a curve using the Type 2 curve fitting procedure which yielded the results shown in column 3.

because there is a linear relationship between concentration and peak height and area. Because with HPLC or GLC the standard curve is linear and has a zero intercept (ie zero concentration = zero response) it is sometimes possible to get perfectly acceptable clinical results with single point calibration. Since running standards and controls on an HPLC is very time consuming, single point calibration can speed up assay times considerably. In a clinical laboratory it is important to generate rapid results.

Internal quality control

Internal quality control samples are samples of known value which are assayed alongside the clinical samples. The results of a batch of assays are only acceptable if the results of the internal control or controls fall within an acceptable range (often plus or minus 10% of its target value) previously agreed for the procedure. Internal control samples for analytes are available commercially, but often they are prepared 'in-house'. In the latter case it is very important that they are prepared separately from the calibrators (working standards), and preferably should be prepared by someone other than the person who prepares calibrators. The observations made with regard to the preparation of calibrators apply equally to the preparation of controls. Commercial controls are available for most aminoglycosides, vancomycin and chloramphenicol. They often contain a preservative such as sodium azide and will usually contain more than one drug; both of these situations render them unsuitable for use with microbiological assay. Before using a commercial control with a commercial assay it would be wise to consult the assay manufacturer as to their opinion on the suitability of the choice. Assays where there is not a simple arithmetic relationship between response and concentration (in reality this means all assays except HPLC or GLC) need to be controlled at least at three levels: the *lower*, the *middle* and the *upper* part of the standard curve.

These so-called high, medium and low controls should be run in rotation to ensure the validity of all parts of the standard curve. Running three controls is especially important for those immunoassays which require infrequent calibration. It is quite possible that an immunoassay standard curve could become more and more inaccurate between, say, 8 and 16mg/l and yet remain perfectly acceptable in the range below 8mg/l. A single control of 6mg/l would thus be inappropriate for controlling such an assay.

The results obtained with each internal control should be plotted on a performance chart which should be kept near the assay equipment. One type of graph will be date and time along the abscissa and result (mg/l) along the ordinate. Lines representing acceptable results (eg target value, target plus 10%, target minus 10%) should be drawn in. Each result is plotted on the chart as it is obtained. A rapid visual idea as to accuracy and precision can be obtained from such a chart. When controls go out of specification the assay should be recalibrated and the batch repeated. A second type of plot which is quite useful is a CUSUM plot. The abscissa is the cumulative target value for the control, the ordinate is the cumulative actual result for the control. Over a period of time the CUSUM of the target and the CUSUM of the results should be very similar if the assay has good accuracy. A CUSUM plot is a good way to identify assay bias. Points at which the assay is recalibrated should be recorded on plots of control results.

External quality assessment

Worldwide there are several schemes which address themselves to the external quality assessment (EQA) of antibiotic assays. These schemes distribute samples for assay to participating laboratories at regular intervals. The laboratories return their results and their performance is assessed statistically, usually by looking at a series of returned results for that laboratory, as well as the single result in isolation, and by making comparisons with results from other laboratories. Examples of such schemes are those run by the College of American Pathologists and INSTAND in Germany. In the UK there is the UKNEQAS Sub-scheme for Antibiotic Assays. This scheme will be explained in some detail since it exemplifies the essence of EQA.

The scheme began in 1973 when it distributed five samples for gentamicin assay to 19 laboratories. In 1980 tobramycin was added, in 1982 netilmicin and chloramphenicol, in 1984 vancomycin, in 1989 flucytosine and in 1991 amikacin.

The scheme currently covers these seven antibiotics and sends samples to over 300 participating laboratories in the UK and overseas. Participating laboratories indicate which analytes they wish to receive and once every month they are sent the desired EQA samples in the form of a single sample of each analyte. Each EQA sample comprises a tube of pooled human serum (1ml) containing a measured concentration of a single antibiotic. Batches of serum are spiked to the

Antibiotic assays

predetermined concentration with the desired antibiotic and then dispensed. These samples are prepared in appropriate numbers on the day of dispatch and are not frozen beforehand. A proportion, however, are held in storage at −20°C for referral purposes. Each sample is labelled with the following information: Antibiotic; Distribution Number; Expiry Date; Return Date.

Laboratories are given two weeks from the day the samples are sent out to perform the assays and return the results along with details of the method used to obtain the results and, for some but not all methods, greater detail in the form of a sub-method. For example, methods used to assay gentamicin include the Abbott TDX, Ames TDA, Gram-negative microbiological assay, Syva EMIT, and the Technicon immunoassay. The methods TDA, TDX and EMIT are divided into sub-methods. For example, the sub-methods of EMIT currently recognized by the sub-scheme are Q (QST or QStat), WM (Wet EMIT assays done manually), WA (Wet EMIT assays done on an autocarousel), AA (Wet EMIT assays done on an autoanalyser such as a RA500, Cobas Bio, Encore etc) and ECP (Special ECP packs designed for the Cobas Mira).

All returns are entered into a computer and each month the following information is calculated for each antibiotic:

N	the number of returns
ALTM	the overall trimmed mean* for all returns
ALTMSD	the standard deviation of the ALTM
ALTMCV	the coefficient of variation

For each method the following are calculated:

MN	the number of returns using that method
MALTM	the overall method trimmed mean
MALTMSD	the standard deviation of the MALTM
MALTMCV	the coefficient of variation

and for each sub-method:

SMN	the number of returns using that sub-method
SMALTM	the overall sub-method trimmed mean
SMALTMSD	the standard deviation of the SMALTM
SMALTMCV	the coefficient of variation.

*Recalculated mean after removal of outliers > two standard deviations from the mean.

Quality control in the microbiology laboratory

Next, each laboratory has its performance for each antibiotic it takes calculated as follows:

%ERR	(Submitted result – target value)/target value × 100
MEAN%ERR	The mean %ERR over this and the previous five distributions
MEAN%ERRSD	The standard deviation of the MEAN%ERR
MEAN+2SD	The modulus of MEAN%ERR plus 2 × MEAN%ERRSD
GROUP	A value derived from MEAN+2SD (See Table 5.2)
SCORE	A value derived from GROUP (See Table 5.2)

This information is compiled into a computer-generated monthly report which is sent to each laboratory with the next distribution of specimens. Laboratories submitting a result >30% from the trimmed mean for any particular analyte are sent a repeat specimen. Every report contains the overall and the method-specific statistics. In addition there is the laboratory-specific performance assessment and the statistical data for the particular sub-method used by the returning laboratory.

Table 5.2 Look-up table for assessing laboratory performance based on MEAN+2SD

MEAN+2SD	GROUP	SCORE	COMMENT
0–20	1	+2	acceptable
>20–30	2	+2	acceptable
>30–40	3	+1	borderline
>40–50	4	+1	borderline
>50–60	5	0	poor
>60–70	6	0	poor
>70–80	7	0	poor
>80–90	8	0	poor
>90–100	9	0	poor
>100–200	10	−1	very poor
>200	11	−1	very poor

MEAN+2SD is the modulus of the mean percentage error obtained over six monthly returns plus two standard deviations above the mean.

Antibiotic assays

The monthly information supplied by the NEQAS scheme (see Fig. 5.1) can assist laboratories in their quality control in a number of ways.

1   Laboratories should look at their return, note their % error and compare their result with the ALTM and the MALTM and SMALTM. If, for example, one month they have a large % error but find that their result compares well with the SMALTM (or MALTM) but not the ALTM then this indicates that their problem is not laboratory specific but sub-method (or method) specific. If their results are consistently different from the SMALTM this suggests a laboratory-specific problem the source of which can be sought. The most common reason for an unexpected very poor result EQA is failing to match the result to the correct specimen. This of course is not a failing of the method used but a human error and indicates that laboratory procedures and protocols may need tightening.

2   They should look at their MEAN+2SD, GROUP and SCORE, note whether it is acceptable or not and compare it with that on the previous report form to determine whether their performance is static, improving or worsening.

3   If they are not content with the results they are returning they can look at the data for alternative methods with regard to popularity (number of returns), accuracy (difference between MALTM and TARGET) and reproducibility (CV%). This will assist them in decisions regarding any change of method. Information on sub-methods is available from the Sub-scheme organiser.

Every six months the UKNEQAS Microbiology Panel meets to discuss performance. Laboratories which are performing sub-optimally (performance in Groups 5 to 11) are sent letters by either the Sub-scheme organiser, the UKNEQAS Panel or its Chairman depending on the seriousness of the performance problem, pointing out the situation.

Currently participation in EQA schemes is voluntary in the UK. However, with moves towards a system of laboratory accreditation this situation may change in the future.

In theory all laboratories should process EQA samples received from recognised schemes as if they were normal clinical samples, but in practice it is often difficult or impossible to do. Laboratories may in the future find themselves being subject to covert EQA. Covertly submitted EQA samples provide a slightly more stringent test for the

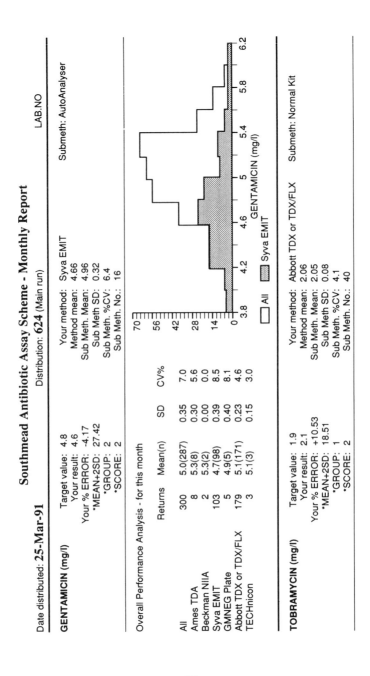

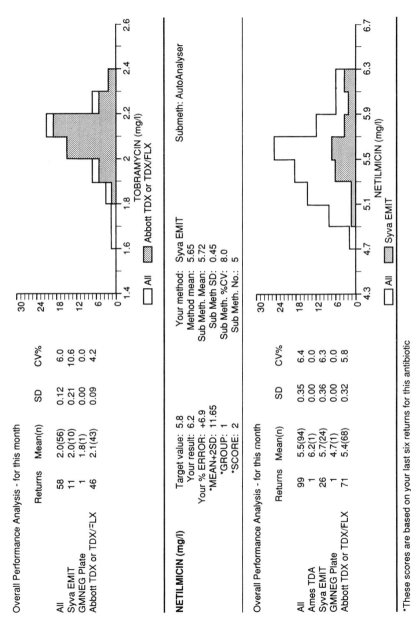

Figure 5.1 A typical NEQAS Sub-scheme for Antibiotic Assays Monthly Report

laboratory staff. Covert EQA has two forms. The first form, which is at present rare, consists of the generation of apparently genuine clinical samples by the laboratory's own director, while in fact they are controls of known value. The second form is practised by the customer to evaluate one laboratory against another. The requesting customer will split samples and send half to one laboratory and half to another. He will then compare the results. Laboratory personnel should appreciate that this is a perfectly acceptable practice. The care and attention that goes into every assay should be as great as that which goes into assaying EQA samples. The laboratory worker should assay every sample assuming his or her results may be compared with those of someone else. As customers are increasingly being asked to pay for assays, they have a right to expect a high quality service.

References

1 European Committee for Standardization (1990). Requirements for labelling of in-vitro diagnostic reagents for professional use. Draft document prEN 375 July 1990, 3–5

2 Perkins A. Statistics of plate assays. In Reeves DS, Phillips I, Williams JD, Wise R. Laboratory methods in antimicrobial chemotherapy. Edinburgh: Churchill Livingstone, 1978; 157–61.

6 Anaerobic bacteriology

KD Phillips

Quality control of anaerobic bacteriology in clinical microbiology is concerned primarily with monitoring the performance of anaerobic culture, and with assessing the suitability and efficiency of the culture media used for isolation and propagation of a wide range of clinically-relevant anaerobes.

From clinical material, the laboratory must be capable of isolating a range of anaerobic bacteria (Table 6.1) varying in oxygen tolerance from the strict anaerobes of the genus *Clostridium* to the facultative but preferentially anaerobic species of *Actinomyces*. Anaerobic bacteria differ widely in their sensitivity to oxygen, but the anaerobes commonly implicated in infectious processes in man are not very sensitive

Table 6.1 Some anaerobic genera of clinical importance

Gram-positive anaerobic bacilli: *Clostridium* *Actinomyces** *Propionibacterium** *Bifidobacterium* *Eubacterium* *Lactobacillus**	Gram-negative anaerobic bacilli: *Bacteroides* *Fusobacterium*
Gram-positive anaerobic cocci: *Peptostreptococcus* *Peptococcus*	Gram-negative anaerobic cocci: *Veillonella*

*Facultative, but preferentially anaerobic

Quality control in the microbiology laboratory

to oxygen and are usually present in clinical specimens in large numbers.

If careful attention is paid to the operation of the anaerobic culture equipment and to growing strict but not unduly fastidious control organisms on appropriate solidified culture media, the isolation of almost all clinically significant anaerobes can be relied upon.

Control of anaerobic culture equipment

Two methods suitable for use in clinical laboratories for the surface culture of obligate anaerobes have been developed using either anaerobic jars or anaerobic cabinets.

Anaerobic jar methods

All modern anaerobic jars employ a room-temperature palladium catalyst to remove oxygen in the presence of excess hydrogen. Various designs of anaerobic jars are commercially available: well-designed jars are manufactured in the UK by Don Whitley Scientific Ltd and by Oxoid Ltd. Guidance on the use of anaerobic jars is provided by Willis,[1] and Willis and Phillips.[2] Jars may be operated by evacuation–replacement methods or by using internal hydrogen–carbon dioxide generators (Oxoid Gas Generating Kit®; Beckton Dickinson GasPak Generator®).

Manometric control of evacuation–replacement operation

Vented jars are essential in this technique for evacuation, replacement and manometry of internal pressure changes. Schraeder valves with chuck attachments are convenient and reliable for these operations and are superior to needle valves.[3]

A partial vacuum, drawn in the jar by a pump, is replaced from a cylinder of compressed gases by a mixture of either 90% hydrogen and 10% carbon dioxide, or 10% hydrogen, 10% carbon dioxide and 80% nitrogen; the latter contains a concentration of hydrogen which is non-explosive in any proportions with air. The degree of partial vacuum that must be drawn initially in the jar is dictated by the concentration of hydrogen in the gas mixture used. Thus, the higher hydrogen concentration requires a preliminary evacuation of about 30cm mercury for satisfactory anaerobiosis to develop; the lower concentration requires an evacuation of at least 60cm mercury. After replacement of the primary vacuum with mixed gases, the initial proportions of hydrogen and oxygen should be approximately 2:1 by volume so that the maximum secondary vacuum (indicated manometrically) will develop in the presence of active palladium catalyst. The development of a substantial secondary vacuum (20–30cm mercury for 90% hydrogen mixture; 5–10cm for 10% hydrogen mixture) is important as a reliable indication that the jar is airtight and that an adequate quantity (at least 1g/litre jar volume) of active catalyst is present. Catalysis should then be allowed to proceed for 15–20 minutes, during which time a secondary vacuum will develop if the jar is functioning efficiently. The pressure is equilibriated by adding further gas mixture and the jar can be incubated. The importance of manometry of pressure changes occurring within the jar cannot be over-emphasized, since jar integrity and adequate catalyst activity are critical to the development and maintenance of anaerobiosis.

Check airtightness and catalyst function routinely before incubation. If these tests are neglected, there will inevitably be failures of anaerobiosis. With transparent polycarbonate jars, airtightness can be demonstrated by introducing a small volume of water into the empty jar, applying a partial vacuum and inverting the jar to cover completely the vents and O-ring seal with water. Any inward leakage of air at the valves or seal will be instantly apparent by bubble formation.

Control of internal gas generators

Hydrogen–carbon dioxide-generating envelopes are efficient means of creating anaerobiosis in sealed jars. However, the cycle of pressure changes during catalysis is slow and more variable than that obtained with evacuation–replacement techniques.[4,5] One cannot control the efficiency of the system by pre-incubation manometry, therefore chemical and biological indicators of anaerobiosis must be used instead.

Anaerobic cabinets

Although simple anaerobic jar techniques can be entirely effective for the recovery of anaerobes from clinical specimens, commercially-available anaerobic cabinets are increasingly used in clinical laboratory practice. An anaerobic cabinet equipped with glove ports and an airlock for transfer of materials provides an oxygen-free environment

Quality control in the microbiology laboratory

in which conventional bacteriological techniques can be applied to the isolation and culture of obligate anaerobes in conditions of strict and continuous anaerobiosis. Several designs of anaerobic cabinet are available commercially in the UK; the Whitley models (Don Whitley Scientific) and the Microflow Anaerobic System® (MDH Ltd) are highly recommended.[6,7,8]

Anaerobiosis in cabinets is achieved with standard palladium catalyst pellets and a non-explosive gas mixture of 10% hydrogen, 10% carbon dioxide and 80% nitrogen, maintained at a slight positive pressure. Gas leaks may develop at many points in this type of complex equipment and must be detected and rectified as soon as possible. Leakage through interchange door and port gaskets is a common problem; unless the seals are tight too much gas is used, leading to rapid exhaustion of the cylinders. This is not only expensive, but may lead to failure of anaerobiosis due to lack of gas. A highly sensitive, portable electronic gas leak detector is marketed by Don Whitley Scientific and is essential for users of anaerobic cabinets. Control of anaerobiosis in cabinets is achieved by continuous monitoring of the atmosphere by means of chemical and biological indicators.

Carbon dioxide

Carbon dioxide is essential or stimulatory for most anaerobic bacteria and is a well-established supplement to anaerobic atmospheres at a concentration of between 5% and 10%.[9,10,11] The precise concentration is probably not critical for growth and isolation of anaerobes but may assume great importance when well-controlled atmospheres are required, as for example in antibiotic sensitivity testing.

A carbon dioxide concentration of 10% by volume in the cylinder is commonly employed. However, in anaerobic jars operated by the evacuation–replacement method, it is worth noting that unless there is virtually complete initial evacuation, the actual concentration of carbon dioxide achieved is less than 10%. For example, a jar evacuated to 30cm mercury will have, after replacement of the primary and secondary vacuums with a gas mixture of 90% hydrogen and 10% carbon dioxide, an atmosphere containing about 7.5% carbon dioxide.

The carbon dioxide content of the atmosphere in jars operated with gas generators may be more variable,[12] and depends on the degree of dissolution of the bicarbonate tablet contained in the envelope. The concentration of carbon dioxide in an anaerobic cabinet atmosphere is, however, directly determined by the concentration in the gas supplied from the mixed gas cylinder, and its consistency is assured by the manufacturer's quality control analyses.

Catalyst

The optimal amount of catalyst pellets in anaerobic jars is 1g/litre jar volume; this proportion is now widely accepted for use in all modern anaerobic jars.[13] Palladium coated alumina pellets are inactivated by moisture (but can be reactivated by heating at 160°C for 90 minutes), or they may be progressively and irreversibly poisoned by products of bacterial metabolism, especially hydrogen sulphide. Failure of the catalyst, whether from moisture inactivation or poisoning, is sudden in that activity can be lost between consecutive cycles of operation. For evacuation–replacement methods, it is important to monitor catalyst activity each time the jar is set up. If the jar is not leaking and if the expected degree of secondary vacuum is not achieved, the catalyst is at fault and must be changed. No definite figure can be placed on the life of an individual catalyst unit. A crude indication of catalytic activity may be provided either by warming of the jar lid adjacent to the catalyst sachet or, in transparent polycarbonate jars, by the mist of condensation formed on the wall of the jar. These cursory observations do not in themselves guarantee that the system is airtight, nor do they imply that catalysis is adequate. I highly recommend a catalyst sachet, marketed by Oxoid Ltd, which contains palladium catalyst pellets wrapped in perforated aluminium foil and encased in wire gauze. The aluminium foil protects the friable catalyst pellets from physical damage and also acts as a heat sink during the exothermic reaction between hydrogen and oxygen, thus minimizing the risk of ignition of gases if high hydrogen concentrations are employed.

In anaerobic cabinets, the catalyst unit (500g) is largely protected from moisture inactivation by a dehumidification device either in the form of silica gel, or by a refrigerated condenser. Minimize poisoning of the catalyst by volatile end-products of anaerobic metabolism by the use of an atmospheric scrubber (Anotox® from Don Whitley Scientific) containing activated charcoal.[14] Nevertheless, the catalyst

Quality control in the microbiology laboratory

should be replaced twice yearly; this is catered for in the service contracts offered by cabinet manufacturers.

Indicators of anaerobiosis

With all types of anaerobic culture equipment, an indicator to verify the development and maintenance of anaerobic conditions is essential. Indicators may be either chemical or biological.

Chemical indicators   The chemical indicators commonly employed are the redox dyes methylene blue or resazurin. For anaerobic jars, convenient disposable methylene blue indicator strips are available commercially from Beckton Dickinson; the more sensitive resazurin strips are obtainable from Oxoid.

It is important to recognize, however, that chemical (and biological) indicators only indicate failure after incubation. When used with evacuation–replacement anaerobic jar methods, they should be considered as adjuncts to the essential prospective control afforded by pre-incubation manometry.

For anaerobic cabinets, a resazurin indicator solution is an effective means of monitoring the maintenance of anaerobic conditions. Reduced resazurin is much more sensitive to traces of atmospheric oxygen than methylene blue. A formula for a suitable resazurin solution is given in Appendix 6.1. Before use, a 50ml aliquot is steamed briefly and placed in the cabinet. This solution should become colourless within a few hours and remain colourless if anaerobiosis is effective. Occasionally during work periods, a slight pink tinge may develop which should soon revert to colourless; this is acceptable atmospheric variation and often occurs when large numbers of plates are transferred in or out of the chamber. Replace the indicator solution described at weekly intervals.

Biological indicators   Anaerobic conditions in all types of culture equipment should be monitored biologically; failure of a moderately-exacting anaerobe such as *C tetani* to grow on the surface of an adequately-nutritious solidified medium indicates a failure to achieve satisfactory anaerobiosis. In practice, if *C tetani* grows, all other clinically-relevant species of obligately and preferentially anaerobic bacteria should be recovered.

The use of *Pseudomonas aeruginosa* as a biological indicator is not recommended because it may occasionally grow in conditions of strict

Anaerobic bacteriology

anaerobiosis as a result of anaerobic respiration coupled to undefined medium ingredients.

By far the best biological indicator system is provided by growth of a control strain of *Clostridium perfringens* on a blood agar plate on which is placed a metronidazole disc (5μg). In conditions of efficient anaerobiosis this relatively undemanding anaerobe will grow and be fully sensitive to metronidazole; if traces of oxygen are present, growth may still occur but the culture will appear resistant to the drug.[15] A suitable strain of *Clostridium perfringens* can be most satisfactorily maintained for control purposes as a spore suspension in 50% ethanol in saline; it will remain viable for up to a year at room temperature without risk of contamination. Media are directly inoculated with this spore suspension. Spores can be readily produced using the sporulation medium of Phillips[16] (Appendix 6.2). In addition, an appropriate strain of *C perfringens* may be similarly maintained for use as a control organisms in antibiotic sensitivity testing of anaerobes, either by a breakpoint method or by the standard Stokes technique (disc diffusion).

Control of anaerobic culture media

The general methods of preparation of media for the culture of anaerobes are the same as for other bacteriological media. It is necessary, however, to use a rich solidified medium incubated in an efficient and controlled anaerobic atmosphere to ensure representative growth of all the pathogenic anaerobes that may be present in a clinical specimen.

The best anaerobic growth is obtained on freshly-prepared media; however, plates for general purposes may be stored aerobically with little detriment for up to one week. Occasionally, pre-reduction of media by storage in an anaerobic atmosphere facilitates the growth of some exacting species. Whenever subcultures are made from a culture of an anaerobe, aerobic control plates should always be inoculated in parallel; these should remain sterile if aerobic contaminants are absent.

A variety of antimicrobial agents may be incorporated in general purpose media for the selective isolation of particular anaerobic species. Among the most generally useful selective agents for clinically-relevant anaerobes are neomycin and neomycin/vancomycin. A

Quality control in the microbiology laboratory

special selective medium containing cyloserine and cefoxitin is used for the isolation of *C difficile*.[17,18] For further information on the use of selective agents in anaerobic bacteriology see Wren,[19] Willis and Phillips[2] and Sutter *et al*.[20] Selective media should always be used in parallel with unselective solidified media; selective media are never perfect in that they frequently inhibit to some degree the growth of organisms whose selection is required. Moreover, resistant strains of species which the medium is designed to suppress are by no means uncommon. The use of discs containing 5μg metronidazole is a valuable aid for distinguishing between colonies of obligate and facultative anaerobes on both selective and unselective agar media; obligate anaerobes are all sensitive to metronidazole in a strictly anaerobic atmosphere.

All batches of both selective and unselective media should be quality controlled using stock strains of anaerobes representative of those organisms commonly encountered or sought in clinical specimens. The range need not necessarily be wide; suggested species include *C perfringens*, *C tetani*, *C difficile*, *Bacteroides fragilis*, *Fusobacterium necrophorum*, *Peptostreptococcus anaerobius*. Although appropriate control organisms can be obtained from the National Collection of Type Cultures, fresh clinical isolates may be preferable; if deemed necessary, identification can be confirmed by the Anaerobe Reference Unit, Cardiff Public Health Laboratory. Selective media must be controlled by demonstrating adequate growth of obligate anaerobes, and by the suppression under anaerobic conditions of appropriate facultatively anaerobic control species. Such organisms include *E coli*, *Proteus spp*, *S aureus* and *Streptococcus pyogenes*.

For quality control of both media and anaerobiosis, stock strains of clostridia are conveniently maintained as spore suspensions in 50% ethanol in saline. Non-sporing anaerobes and facultative anaerobes are best stored as heavy suspensions at −70°C in brain heart infusion broth, to which 10% glycerol is added as a cryoprotectant.

References

1 Willis AT. Anaerobic bacteriology; clinical and laboratory practice, 3rd edn. London: Butterworth, 1977.
2 Willis AT, Phillips KD. Anaerobic infections: clinical and laboratory practice. London: Public Health Laboratory Service, 1988.
3 Burt R, Phillips KD. A new anaerobic jar. *J Clin Pathol* 1977; **30**: 1082–4.

Anaerobic bacteriology

4 Collee JG, Watt B, Fowler EB, Brown R. An evaluation of the GasPak system in the culture of anaerobic bacteria. *J Appl Bacteriol* 1972; **35**: 71–82.

5 Seip WF, Evans GL. Atmospheric analysis and redox potentials in the GasPak system. *J Clin Microbiol* 1980; **11**: 226–33.

6 Brazier JS. An evaluation of the MDH Microflow Anaerobic System. *Med Technol* 1985; **Oct**: 8–9.

7 Phillips KD, Willis AT. Appraisal in the diagnostic laboratory of three commercially available anaerobic cabinets. *J Clin Pathol* 1981; **34**: 1110–13.

8 Sisson PR, Ingham HR, Byrne PO. Wise anaerobic work station: an evaluation. *J Clin Pathol* 1987; **40**: 286–91.

9 Reilly S. The carbon dioxide requirements of anaerobic bacteria. *J Med Microbiol* 1980; **13**: 573–9.

10 Stalons DR, Thornsberry C, Dowell VR. Effect of culture medium and carbon dioxide concentration on growth of anaerobic bacteria commonly encountered in clinical specimens. *Appl Microbiol* 1974; **27**: 1098–104.

11 Watt B. The influence of carbon dioxide on the growth of obligate and facultative anaerobes on solid media. *J Med Microbiol* 1973; **6**: 307–14.

12 Ferguson IR, Phillips KD, Tearle PV. An evaluation of the carbon dioxide component in the GasPak anaerobic system. *J Appl Bacteriol* 1975; **39**: 167–73.

13 Watt B, Hoare MV, Collee JG. Some variables affecting recovery of anaerobic bacteria: a quantitative study. *J Med Microbiol* 1973; **77**: 477–54.

14 Brazier JS. Appraisal of Anotox, a new anaerobic atmospheric detoxifying agent for use in anaerobic cabinets. *J Clin Pathol* 1982; **35**: 233–38.

15 Milne SE, Stokes EJ, Waterworth PM. Incomplete anaerobiosis as a cause of metronidazole 'resistance'. *J Clin Pathol* 1978; **31**: 933–5.

16 Phillips KD. A sporulation medium for *Clostridium perfringens. Lett Appl Microbiol* 1986; **3**: 77–9.

17 George WL. Sutter VL, Citron D, Finegold SM. Selective and differential medium for isolation of *Clostridium difficile. J Clin Microbiol* 1979; **9**: 214–19.

18 Phillips KD, Rogers PA. Rapid detection and presumptive identification of *Clostridium difficile* by p-cresol production on a selective medium. *J Clin Pathol* 1981; **34**: 642–4.

19 Wren MWD. Multiple selective media for the isolation of anaerobic bacteria from clinical specimens. *J Clin Pathol* 1980; **33**: 61–5.

20 Sutter VL, Citron DM, Edelstein MAC, Finegold SM. Wadsworth anaerobic bacteriology manual, 4th edn. Belmont, California: Starr, 1985.

Quality control in the microbiology laboratory

Appendix 6.1   Resazurin indicator solution

The solution may be prepared by dissolving 100g Tris and 5g glucose in 500ml distilled water. To this is added 25ml 0.1% solution of resazurin in water. The complete indicator solution is dispensed in 50ml amounts in screw-capped bottles. At this stage the resazurin indicator is blue and must be converted through pink to colourless by gentle steaming. The solution will deteriorate rapidly if subjected to prolonged heating.

Appendix 6.2   A sporulation medium for *Clostridium perfringens*

The sporulation medium is prepared by dissolving 39.5g Oxoid blood agar base no 2 and 10g desiccated ox bile (Oxoid) in 1 litre distilled water prior to sterilization by autoclaving at 121°C for 15 minutes. When cooled to about 50°C, 5g sodium bicarbonate and 0.5ml quinoline (BDH Chemicals Ltd, product no 30012) is added and mixed to dissolve completely. Defibrinated horse blood (50ml) is added and the sporulation medium is dispensed in 25ml amounts in plastic petri dishes. The pH of the medium is 8.5. The critical components of the medium for inducing sporulation in this species are bile, bicarbonate and quinoline.[16]

7 Preservation of control strains

JJS Snell

Every laboratory needs to maintain a collection of the strains used in quality control procedures. These strains form the basis of a culture collection which can be usefully supplemented with other strains for use in teaching and research. Cultures maintained in a culture collection must remain viable and pure and must retain important characteristics. It is not easy to ensure compliance with these requirements, as cultures have apparently limitless potential for becoming contaminated, altering their characteristics or dying. Suitable preservation techniques can be used to minimize these events.

A variety of preservation methods, many of which were designed for particular groups of microorganisms, exist. Several reviews of these methods and their applications are available.[1,2,3] This chapter is limited to a discussion of the factors to be considered in selecting a preservation method, and a brief description of some commonly-used methods. Technical details of these methods are fully described elsewhere.[4] Training courses on preservation of microbes are organized regularly by the United Kingdom Federation for Culture Collections and details of these are announced in the meetings sections of the scientific press.

Choice of a maintenance method

There is no universally-applicable method that will successfully preserve all microorganisms and meet the logistic requirements of all users. The method of choice depends on the balance of advantages and disadvantages and the particular circumstances of the user.

Quality control in the microbiology laboratory

Factors to be considered in the choice of a maintenance method are discussed below.

Survival periods

Cell death may occur both in the preservation process and during subsequent storage. To avoid the necessity of frequently re-preserving cultures, the method chosen should minimize such loss of cells.

Stability of characters

Whether strains are preserved because they exhibit a particular property (such as sensitivity to an antibiotic), or for other reasons, it is desirable that characteristics remain stable during storage. There are two mechanisms by which changes can occur. Selection of a minority population of cells can occur through the death of a proportion of cells, either during the preservation process or on storage; such reductions in the number of viable cells may appear unimportant to the user if a high initial concentration of cells is used. However, such selection can result in changes in the properties of the culture over a period of time and preservation methods should minimize cell death. Another mechanism of change is through mutation or loss of plasmids during preservation and storage and the method chosen should minimize this.

Purity

Cultures preserved for most applications should remain pure, or certainly of known composition, and the preservation method should minimize the chances of contamination.

Expense

The cost of maintaining cultures includes the cost of staff time, equipment, materials and general factors such as storage space and power supplies. The high capital costs of equipment for some methods, such as liquid nitrogen storage and freeze-drying, may be offset against the labour savings resulting from the long-term stability of cultures preserved by these methods. Costs of preservation must also be balanced against the value of a culture in terms of the amount of work spent in the selection and characterization of suitable strains.

Number of cultures

The amount of operator time required for initial preservation and subsequent manipulation and the storage space required may limit the number of cultures that can be conveniently maintained. A method suitable for preservation of a small collection may prove too labour intensive when the number of strains increases.

Supply and transportation of cultures

If cultures are to be distributed to other workers, replicates of the cultures will be needed. These may be prepared as required or produced in bulk and stored for later distribution. The convenience of either approach depends on the method in use, the number of cultures and the frequency of distribution.

Frequency of use of cultures

Cultures used for quality control purposes may be used frequently within a laboratory; ease of resuscitation, stability of properties and the dangers of contamination of stock cultures through frequent access need to be considered.

Methods

Sub-culture

Although this is probably the most widely used method of preservation, it has many disadvantages. A suitable medium is inoculated, incubated at an appropriate temperature to obtain growth and stored under suitable conditions. The process must be repeated at intervals to ensure the preparation of a fresh culture before death of the old one. The safe time interval between sub-cultures depends on medium, storage conditions and the particular organism. Many bacteria such as staphylococci and coliforms will survive for several years under suitable conditions, whereas others, such as *Neisseria* spp, may require subculture after only a few weeks.

This method is inexpensive in equipment but is labour intensive if large numbers of organisms requiring frequent sub-culture are kept. Cultures are easily resuscitated, since a single sub-culture results in an active culture. The method is technically simple and is applicable to a wide range of microorganisms. Contamination is a major prob-

Quality control in the microbiology laboratory

lem and the risk is present at each sub-culture. Apart from the undesirability of mixed cultures, contaminants may overgrow and kill the original culture. Contamination may be reduced by sound microbiological technique and the pre-incubation of media before use. A two-tube system, where one tube is kept as a stock and the other as a working culture, reduces the risk of contaminating the stock strain through frequent manipulation.

The risk of mislabelling or transposition of cultures is high, particularly where strains require frequent sub-culture. Numbers written on containers may be misread and often become completely altered after several sub-cultures. Avoidance of operator fatigue and the use of typewritten numbers can help to overcome these problems.

Loss of viability is a constant problem with this method. If strains with different survival characteristics are maintained, a protocol is required to ensure the timely sub-culture of strains before they lose viability.[5]

Sporadic loss of cultures may occur because of dehydration of the media through an imperfect container seal. Plastic screw-caps are particularly prone to this. Whole batches may be lost because of faulty or contaminated media.

Alterations in strain characters are frequent with this method and increase with the frequency of sub-culture. Distribution of cultures preserved by this method to other workers is not particularly convenient as sub-cultures must be prepared and checked for purity, and survival in transit is not guaranteed.

Survival times may be prolonged with the use of suitable media and a great many media have been described for this purpose.[1] The tendency is to use unenriched, nutritionally-limited media without carbohydrates as the acid produced can kill the culture.

Storage periods can be extended by reducing the metabolic rate of the organisms, by either restricting the availability of air by overlaying the culture with liquid paraffin, or storing at 5°C. However, not all microorganisms survive well at lower temperatures; *Neisseria* spp, for example, appear to survive best at 35°C.

Freeze-drying

Freeze-drying has been widely used to preserve yeasts, fungi, bacteria and some viruses. Centrifugal freeze-drying is most widely used; this is a process by which cultures are frozen by evaporation, dried over a refrigerated condenser or phosphorus pentoxide and sealed under vacuum or an inert gas in glass ampoules.

The advantages of freeze-drying are: long-term maintenance of viability (periods of fifty years or more have been recorded with some organisms); relative stability of characters during storage; undemanding storage requirements; suitability for batch production and ease of distribution. For these reasons freeze-drying is the method of choice for service culture collections.

The method does have some disadvantages: the capital cost of equipment is high and regular maintenance is essential; the method is fairly labour intensive and there are limits to the numbers of cultures that can be processed in a day. However, large batches of single cultures can be prepared and the labour costs per ampoule may be low. Stability of characters during storage is generally good, although genetic changes and loss of plasmids may occur. Population changes through selection are a greater hazard since, with some species, drops in viability of a thousand-fold during processing are not uncommon. For the user, the sealed glass vials are time-consuming to open and appropriate safety procedures must be followed.

Drying on gelatin discs

In this method, organisms are suspended in a nutrient gelatin medium, drops of which are dried before storage over silica gel in screw-capped bottles. The drops dry to form flat, discrete discs, each one of which provides material for one sub-culture, achieved by rehydrating in warm broth and plating. Although in the original method drops were simply dried in a desiccator,[6] pre-freezing of the drops and subsequent drying under vacuum (a freeze-dryer may be used for this purpose) is a convenient alternative. A variety of bacteria have been preserved by this method with survival over several years.[7,8,9] This method is well suited to strains which are frequently used (such as quality control strains), since sub-cultures are easily made by removing a gelatin disc. There is little risk of contamination of the remaining stocks since any contaminants introduced during removal of a disc are unable to grow in the dry conditions of storage. The method is labour intensive in preparation and is best suited for use with small numbers of selected strains.

Freezing on glass beads

In this method[10] small glass beads are immersed in cultures suspended in glycerol broth, drained and stored in plastic, screw-capped vials at around −70°C. The method has been successfully applied to a wide range of bacteria. It is very quick and easy and requires no subsequent intervention during storage. Each glass bead provides material for one sub-culture and allows large batches to be stored in minimal space. The method is ideally suited to storage of large in-house collections.

Disadvantages of this method include the high capital cost of a −70°C freezer, which also occupies a lot of space. It is possible to store cultures at temperatures nearer to freezing, from −20°C and below. There are theoretical objections to the use of temperatures around −20°C. In practice, hardier organisms will survive quite well at higher temperatures but the more delicate, which present problems with most preservation methods, survive best at lower temperatures. Provision must be made against breakdown of the freezer with availability of some back-up freezer space. Many organisms appear to survive 'melt-down' caused by freezer failure quite well.[11] The method is not well suited to frequent distribution of cultures to other workers as sub-cultures need to be made before issue.

Anecdotal reports reveal mixed experiences among users of this method: some find it satisfactory for a wide range of the most delicate organisms, while others experience loss of cultures. Like any method, details of technique are probably important and the individual user will need to ascertain that the technique is working well under local circumstances before committing large collections of strains. New batches of beads should be checked for suitability.

Storage in liquid nitrogen

A wide range of microorganisms have been successfully preserved in liquid nitrogen. The major advantage of this method is that although reductions in viability may occur during freezing and warming of the cultures, virtually no loss occurs during storage. The method may be used to preserve many organisms that do not survive freeze-drying. The longevity and stability of preserved cultures make this the method of choice for valuable seed stock material. Storage space may be maximized by combining liquid nitrogen preservation with the use

Preservation of control strains

of glass beads as described above. The method is not labour intensive and requires no further intervention during storage.

The disadvantages of this method include the high capital cost of equipment and the need to ensure a constant supply of liquid nitrogen. There is a risk of explosion if containers are kept in the liquid phase, as liquid nitrogen may leak into containers and expand rapidly when the container is warmed. This can be avoided by storage in the vapour phase (ie above the liquid nitrogen). The method is not very convenient for distribution of cultures to other workers, as sub-cultures need to be prepared.

Quality control of cultures

Preservation of cultures, like any other procedure in microbiology, must be subjected to quality control. Endless time can be spent on such quality control and a large proportion of the work of the service culture collections is devoted to this. The average laboratory will have limited resources to spend on such quality control and must be selective in the procedures adopted. The essential parameters to be controlled are purity of the culture, that the correct culture has been preserved and that the culture retains the characteristics for which it was preserved. Purity should be checked on non-selective media both before and after preservation. Common contaminants often grow poorly at 37°C and plates should be left at room temperature for several days after incubation. That the correct culture has been preserved and retains the desired characters can only be determined by characterization of the strain, using tests selected to provide the desired information.

For a culture collection to be useful, both now and in future, it is essential that strains are properly documented. Data may be stored in notebooks, card files or in computer systems. Useful information to be recorded is as follows:

1 laboratory reference number
2 source
3 reason for accession
4 any relevant properties of the strain
5 method and dates of preservation
6 cultural requirements
7 date of accession.

References

1 Lapage SP, Redway KF. Preservation of bacteria with notes on other microorganisms. Public Health Laboratory Service Monograph no 7. London: Public Health Laboratory Service, 1974.

2 Lapage SP, Redway KF, Rudge R. Preservation of microorganisms. In: Laskin AI, Lechevalier HA, eds: Chemical Rubber Company handbook of microbiology, vol 2. Florida: Chemical Rubber Company Press, 1978: 743–58.

3 Snell JJS. General introduction to maintenance methods. In: Kirsop BE, Snell JJS, eds: Maintenance of microorganisms – a manual of laboratory methods. London: Academic Press, 1984: 11–21.

4 Kirsop BE, Snell JJS, eds. Maintenance of microorganisms – a manual of laboratory methods. London: Academic Press, 1984.

5 Skerman, VBD. The organisation of a small general culture collection. In: Pestana de Castro AF, Da Silva EJ, Skerman VBD, Leveritt WW, eds: Proceedings of the second international conference on culture collections. Brisbane: University of Queensland, 1973: 20–40.

6 Stamp L. The preservation of bacteria by drying. *J Gen Microbiol* 1947; **1**: 251–65.

7 Snell JJS. Maintenance of bacteria in gelatin discs. In: Kirsop BE, Snell JJS, eds: Maintenance of microorganisms – a manual of laboratory methods. London: Academic Press, 1984: 41–5.

8 Obara Y, Yamai S, Nikkawa T, Shimoda Y, Miyamoto Y. Preservation and transportation of bacteria by a simple gelatin disc method. *J Clin Microbiol* 1981; **14**: 61–6.

9 Yamai S, Obara Y, Nikkawa T, Shimoda Y and Miyamoto Y. Preservation of *Neisseria gonorrhoeae* by the gelatin disk method. *Br J Vener Dis* 1979; **55**: 90–3.

10 Jones D, Pell PA, Sneath PHA. Maintenance of bacteria on glass beads at −60°C to −70°C. In: Kirsop BE, Snell JJS, eds: Maintenance of microorganisms – a manual of laboratory methods. London: Academic Press, 1984: 35–40.

11 Pell PA, Sneath PHA. A note on the survival of bacteria in cryoprotectant medium at temperatures above 0°C. *J Appl Bacteriol* 1984; **57**: 165–7.

8 Immunoassays

TG Harrison and A Malic

During the last ten years both the range of infections diagnosed and the number of specimens examined by immunoassay methods have increased greatly. Immunoassays may be used to detect, quantify and identify microbial pathogens or their components or antibodies. Immunoassays based on agglutination, complement fixation, neutralization or immunofluorescence are well established. Recent technological advances (eg in the application of monoclonal antibodies and recombinant proteins, and the development of signal amplification systems) have increased the type and range of immunoassays used in routine diagnostic laboratories.[1,2,3] This has resulted in increasing workloads and a greater reliance on instrumentation and automation.

Stringent quality control (QC) programmes are needed to ensure the accuracy and reliability of test results obtained with both well established and newer assays. Data from the UKNEQAS for medical microbiology has shown that while most laboratories may perform well, in some laboratories QC programmes are not used, or are used ineffectively. Deficiencies in QC procedures thus revealed include the following.

1. Failure to use any QC procedures.
2. Failure to use recommended QC procedures.
3. Failure to analyse and act on the results obtained using QC samples.
4. Deviation from assay protocols.
5. Continued reporting of results although QC samples indicate that the assay run was unacceptable.

The quality assurance of diagnostic immunoassays encompasses any and all measures necessary to ensure that a valid result is obtained. This includes all aspects of the diagnostic service from collection of the specimen to the issue of the laboratory report. However, this contribution focuses mainly on quality control procedures and in particular those that directly influence test performance. The design and evaluation of an immunoassay is a subject outside the scope of this chapter, but it should be recognized that the sensitivity, specificity, bias and precision inherent in the design of a particular test will have a major influence on the reliability, and hence quality, of a test result. It is therefore essential that these parameters have been determined and the test fully validated before clinical decisions are based on results obtained using it.

QC procedures are essentially of two types: those that are undertaken to minimize error in the assay (preventive procedures) and those that are undertaken to verify technique and detect errors which have occurred (verification procedures).

Preventive procedures

Specimen quality

The quality of the clinical specimen is an important determinant of performance so specimens should be collected, transported and stored under specified and controlled conditions. It is often difficult to directly influence procedures that take place prior to the arrival of a specimen in the laboratory, but these can affect the accuracy of the final report (eg the point in the course of an illness when a specimen is collected, the type of transport media used and the time a specimen is in transit). Identification of the reasons for inadequate or inappropriate specimens can result in corrective action. Written protocols together with constant communication between the users and laboratory staff both help to ensure that appropriate specimens are received. Individuals responsible for sending inappropriate specimens rejected by the laboratory should be informed directly and advised so that suitable replacements can be collected. Specimen quality must be maintained after arrival in the laboratory. Where it is necessary to retain specimens, because of a delay in testing or for re-examination at a later date (eg to test acute and convalescent sera in parallel), the conditions of storage must be shown not to affect the test results. Sera can usually be stored at below $-20°C$ for extended periods but they should not be subject to repeated freezing and thawing.

Documentation

Method sheets or protocols should be prepared for in-house immuno-assays, or obtained from the manufacturer for commercial kits. Protocols, which should include a discussion of the principles of the assay, should be collated into a methods manual and made available to all staff members. These protocols should be continually updated (see Introduction).

Accurate records should be kept for all reagents bought, prepared and used in the laboratory. This might, for example, enable a sudden change in assay performance to be traced quickly and easily to a particular batch of a reagent. Not only is the ability to 'trace the source' important in terms of QC, but it is essential in limiting liability as a supplier (or producer) of goods under the 1987 Consumer Protection Act.[4]

Equipment

Diagnostic immunoassays require the use of reliable and accurate equipment. laboratory equipment is becoming increasingly sophisticated, with a trend towards automated and integrated systems.[5,6] Although this allows large numbers of specimens to be processed and the results obtained to be analysed objectively, it places considerable reliance on the correct functioning of the equipment. Neglected or faulty equipment will produce unreliable test results and thus it is essential that all equipment is maintained to a high standard. Some commonly encountered faults (and appropriate control measures) include:

1 Incubator temperature drift – use fan-circulated incubators and stirred waterbaths to avoid local temperature gradients, and monitor the temperature daily.
2 Inaccurate delivery of liquids by pipettes – use only precision pipettes and check their calibration regularly (eg monthly). 'Accuracy check kits' can be obtained from some pipette manufacturers.
3 Inadequate washing of microtitre plates – ensure that washing probes are not blocked with deposits from washing buffers, regularly clean all probes and tubing. After washing a plate, visually check that it is dry.

Quality control in the microbiology laboratory

Particular attention should be paid to the regular servicing and maintenance of spectrophotometers, scintillation counters and other equipment which produces quantitative readings that cannot easily be verified by the operator. Details of servicing and routine checks undertaken on equipment should be recorded so that their performance can be monitored over extended periods of time.

Reagents

Reagents should only be used for the purpose for which they are intended (eg immunofluorescent reagents are not suitable for agglutination tests and *vice versa*). Commercial reagents should be stored at the recommended temperature on receipt and discarded on their expiry date.

Where reagents are prepared in house, stability studies should be undertaken to determine shelf life and optimum storage conditions[7] and they should be discarded on their expiry date. The potency, homogeneity, specificity, sensitivity and purity of each batch prepared must be tested.

Label reagents clearly with their identity, the dates of preparation and expiry, a batch number and any dilution required. Prepare working dilutions only in the appropriate diluents. Examine before use for evidence of microbial contamination or any other sign of deterioration.

The need to replace a reagent should be anticipated well in advance and new batches should be obtained, or prepared, and examined in parallel with the current batch to ensure that they are suitable for use. Where possible, within the constraints of shelf life, it is preferable to buy particular batches of reagents, such as conjugates, in bulk. Their working strength can then be determined by 'chess-board' titration against specimens of known activity and aliquots of the neat reagent can be made and stored for later use. If the producer's recommended storage conditions are changed (eg an aliquot of neat conjugate is diluted to working strength before storage), the aliquot should be subjected to stability studies.

Commercial kits

Many of the immunoassays now used in microbiology laboratories are obtained from commercial manufacturers as kits. Kits comprise the reagents, documentation and appropriate QC materials to allow the entire immunoassay to be performed. Usually the components of a particular kit will have been quality controlled by the manufacturer as a batch and these should not be interchanged either with components of a similar kit (with a different batch number) or with components of another manufacturer's kit. To ensure the results obtained are interpreted as described by the manufacturer the protocols provided must be followed precisely. If the described protocol is altered, the modified test must be considered an entirely different test and fully evaluated as described below for in house assays.

In-house immunoassays

The development, characterization and evaluation of an immunoassay is demanding and time consuming and should only be undertaken where there is a clear need for an assay and no suitable alternative already exists. Immunoassays prepared in house must be rigorously standardized, evaluated and quality controlled before being used routinely. Not only must the whole kit be evaluated, but each component reagent should be subjected to the quality control procedures described in the reagents section. Even if an in-house assay incorporates commercial reagents of known quality and potency it will still need to be fully evaluated.

Standardization of assays is outside the scope of this contribution, but the test performance parameters which should be established for a quantitative immunoassay are shown in Table 8.1. Further details can be found elsewhere.[8,9,10]

Assays which are infrequently undertaken

A high standard of assay performance is most easily obtained when the assay is in regular use. Where a particular immunoassay is infrequently requested difficulties may be encountered in undertaking the assay (eg because reagents l.ave deteriorated or are out of date) and the quality of the results obtained may be poor (eg due to the inexperience of the operator). It is also unlikely that an adequate QC programme can be sustained. Such problems can, to a large degree, be overcome by performing the assay using previously-tested specimens, at regular intervals, irrespective of the number of requests for it. This is, however, unlikely to be cost effective and referral of all such specimens to another laboratory which regularly performs the test may be preferable.

Quality control in the microbiology laboratory

Table 8.1 Assay performance parameters which should be established in the evaluation of a quantitative immunoassay

---

Dose-response curve (signal proportional to analyte concentration)
Preparation of reference material (if no international/national reference preparation available)
Parallelism studies (between reference preparation and patient specimens)
Precision profile (within-batch and between-batch precision, determination of assay detection limits)
Stability studies (of component reagents and completed kit)
Specificity studies (cross-reactions, interference studies)
Sensitivity studies (% true positives found to be positive in the assay)
Comparison with other methods (against the reference method if one exists)
Field studies (to determine the clinical utility and laboratory performance of the assay)
Inter-laboratory studies

Testing of single specimens

There are increasing demands on many laboratories to provide rapid results for single or small numbers of specimens, both during and outside normal working hours. These tests require the same level of QC as do larger batches of specimens. Positive and negative controls and reagent blanks must be included and the results analysed, even though this might appear an extravagant use of reagents.

Verification procedures

Test verification is achieved by the use of reference and control materials. Each immunoassay performed, irrespective of batch size, should include appropriate controls from which an assessment of the test performance may be made. These control materials are also used to monitor the test performance over a period of time. Control materials should comprise a panel of reference preparations, known positive and negative specimens (QC materials) and reagent controls.

Control materials

Quality control materials QC materials are essential for monitoring both the within-batch and between-batch performance of an immunoassay. All immunoassays should include:

Immunoassays

1 A negative control which contains no analyte. This should be the same type of material as the test specimens (ie if neat serum is being examined then the negative control should also be a neat serum).

2 One or more positive controls which contain the analyte of interest. Again these should be the same type of material as the test specimen.

In quantitative assays, a weakly positive specimen (low positive) should be included, which contains the analyte either at the limit of detection of the test, or at the limit of diagnostic significance. Such a control will usually be very sensitive to any change in assay performance.

Control materials should only be used in the assay for which they have been evaluated and should be processed in the same way as are patient specimens. Control samples should be introduced at intervals in large batches of specimens to monitor any within-batch drift of assay performance. Such drift can occur, for example, because of slight differences in incubation times for the first and last specimens. Alternatively, a larger panel can be tested, including positive controls representative of the range of analyte concentrations likely to be encountered in patient specimens. This latter approach is particularly useful where (as is normal) the precision of the assay varies with analyte concentration. Where possible, all control materials should be processed blind to avoid operator bias. It is important to monitor the stability of control material to avoid misinterpretation of assays incorporating unstable controls.

Often QC materials are in short supply and the users may need to prepare their own material. This should only be done where appropriate material is available (or can be prepared) in large enough quantities to be used for an extended period of time. Such material must be tested to exclude the presence of hepatitis B surface antigen and antibodies to HIV.[11]

Reference materials Reference materials or standards are primarily used to calibrate test procedures and thus improve the standardization of immunoassays. Although a particular laboratory's results for an assay may be very reproducible, unless calibrated against a preparation of defined potency they cannot easily be compared with results obtained from other laboratories. Thus reference materials should be

Quality control in the microbiology laboratory

included in each batch of test specimens and used to calibrate the results obtained.

Typically, reference materials are provided by international or national authorities (eg the World Health Organization and the National Institute for Biological Standards and Control) and are of defined activity and known stability. These serve as primary standards for the calibration of working standards in commercial assay kits, or in-house assays.

In-house standards should be the same type of material as the test specimens. As appropriate clinical material is usually in short supply, specimens are often pooled together to provide sufficient material for use over a long period of time. This is satisfactory provided that care is taken not to include atypical specimens in the pool. Where pools are used to standardize tests other than for hepatitis and HIV serology, the individual components should be tested to exclude the presence of hepatitis B surface antigen and antibodies to HIV. To ensure that standards are suitable for use across the entire assay range when diluted, the procedure of parallel line assays should be used.[12] Ideally such material should be examined in another laboratory (eg a reference laboratory) to ensure that there is independent agreement of its suitability. It is important to recalibrate in-house standards against reference material at regular intervals to confirm their continuing suitability.

A range of international and national reference materials is available for a number of immunoassays and these can be used to:

1   Calibrate test procedures and assess the bias of an immunoassay.
2   Calibrate in-house standards prepared for inclusion as controls.
3   Enable the activity of a sample of unknown potency to be expressed in a defined unitage (eg international units – iu).
4   Prepare 'cut off' control material at the limit of clinical or diagnostic significance (eg a 15iu control for Rubella SRH IgG serology).[13]

Comprehensive lists of standards and reference materials and their sources are available.[14,15] Sources of such material relevant to diagnostic clinical microbiology are shown in Table 8.2.

Reference materials should only be used to calibrate assays for which they are intended, as usually those prepared for use in one immunoassay are not applicable in an assay that uses different antigens or evaluates different antibody specificities.

Table 8.2 Examples of available reference materials

Preparation	Status	iu per ampoule	Available from
WHO international standards and reference preparations			
Anti-streptolysin O serum (human)	1st standard 1959	2160	Copenhagen[1]
Anti-toxoplasma serum (human)	2nd standard 1980	2000	Copenhagen
Anti-measles serum (human)	1st reference preparation 1964	10	Copenhagen
Anti-rubella serum (human)	2nd reference preparation 1970	1000	Copenhagen
Hepatitis B immunoglobulin (human)	1st reference preparation 1977	50	Amsterdam[2]
Hepatitis B surface antigen *ad* subtype	1st standard 1985	100	Potters Bar[3]
NIBS[3] British standards			
Anti-measles serum (human)	1st British standard 1974	5	Potters Bar
Anti-rubella serum (human)	2nd British standard 1986	80	Potters Bar
DMRQC reference materials			
Rubella HAI antiserum	(Code 483)	500	DMR[4]

[1]International Laboratory for Biological Standards, Statens Seruminstitut, 80 Amager Boulevard, Copenhagen, Denmark

[2]International Laboratory for Biological Standards, Central Laboratory, Netherlands Red Cross Blood Transfusion Service, Plesmanlaan 125, Amsterdam, Netherlands

[3]International Laboratory for Biological Standards, National Institute for Biological Standards and Control, Potters Bar, Herts, EN6 3QG

[4]Division of Microbiological Reagents, Central Public Health Laboratory, 61 Colindale Avenue, London, NW9 5HT

Reagent controls Reagent controls (or blanks) are those other than the positive and negative controls which help to check the performance of the reagents. Reagent controls should be included for each reagent stage in the assay. For example, if a heterogeneous enzyme immunoassay has four steps as follows:

step A: coating carrier surface with protein
step B: addition of clinical material
step C: addition of enzyme-labelled protein
step D: addition of substrate

then the following reagent controls should be included in the assay: substrate control (A + D), conjugate control (A + C + D) and specimen control (A + B + D). Test diluent should be used to replace the reagents that are omitted. An increase or variation in the readings obtained for any of these controls will enable the source of error to be determined (for example, unstable substrate or over-strong conjugate).

Assessment of test performance

For qualitative assays that are read visually, the pattern of reactivity should be assessed objectively and independently, ideally by more than one individual. Any specimens giving 'borderline' results must be tested again.

It is essential that any error associated with quantitative assays should also be quantified. There are many approaches to this problem and detailed discussions can be found elsewhere.[9,16,17] However, some of the concepts will be briefly outlined here.

The reliability, or accuracy, of a test result will depend on two major classes of measurement error: the systematic bias and random error.

The systematic bias of an assay is a measure of how close the estimated value is to the true value (in practice, the true value is usually not known but a target value can be arrived at by a consensus of expert opinion). Strict attention to the correct calibration of test results, together with the preventive procedures outlined above, will minimize the systematic bias. However, some bias is usually inherent in the design of an assay and this is outside the user's control.

The random error of an assay is measured as 'precision', or the spread of results around the mean which is obtained when a sample is repeatedly assayed. Typically the precision of an assay is different if

Immunoassays

the results to be compared are obtained from tests within a batch (eg on the same microtitre plate) or between batches (on different microtitre plates). These parameters can be estimated by assaying a specimen many times within one batch (within-batch precision) and once per batch many times (between-batch precision). However, this approach is of limited value as it only provides information concerning the precision of the assay for one specimen at one period of time. Clearly, it is more useful to have continually updated estimates of the assay's precision and this is the objective of QC programmes.

The exact choice of the QC programme to be used will depend on many factors, including the number of specimens examined, the range of analyte concentrations to be measured and the clinical significance of the results obtained. Suitable quality control programmes are described in detail elsewhere.[9,17,18] Typically, control specimens are assayed in duplicate on every occasion the test is performed, and the cumulative data used to continually revise the estimates of within- and between-batch precision. The variation in these estimates of precision (expressed as within-batch standard deviation S_w and between-batch standard deviation S_b) provides much information and can be used to:

1 determine whether the results of a particular assay batch are acceptable or not
2 give an estimation of the error associated with the measurement made on a patient sample
3 assess the significance of changes between serial specimens from the same patient
4 establish quality control charts which enable the performance of the assay to be monitored over long periods of time.

Analysis of control results The acceptability of test results from a batch can be judged by analysis of the control results. The exact criteria chosen for the acceptance or rejection of a batch of results can only be decided on the basis of a knowledge of the patient population being examined and the clinical expectations of the assay. Detailed discussion of these factors and the use of control charts to help in the interpretation of QC data is given elsewhere.[9,17,18]

Analysis of results of test specimens Confidence in the validity of individual test results is greatly increased if specimens are tested in replicate (usually duplicate). There are several good reasons for doing

Quality control in the microbiology laboratory

this: firstly, the mean of replicate results is a better estimate of the true value than is a single result. Secondly, erroneous results are often caused by technical problems (eg inadequate washing of part of a microtitre plate because one washing probe is blocked, or failure to add a reagent to one particular well), and a check for consistency between replicates will often identify such errors. Also, limits of acceptable variation can be established with replicated tests.[9,17,18]

Occasionally, errors may occur which are not detected by analyzing test specimen or control results (eg transcription errors). In such cases the intuition, expertise and clinical awareness of laboratory or medical staff is important in assessing the clinical or diagnostic acceptability of such a result. If unexpected results are obtained it is prudent to repeat the test, or confirm the result using a different assay, before a report is made.

Test specificity The reagent specificity of the components used in an immunoassay should have been thoroughly evaluated by commercial or in-house producers before incorporation into the assay and the reagent controls described above will serve to check this for each batch of tests. However, the specificity of the whole assay may be influenced by factors not revealed by such controls. The types of problems that can arise include:

1 Immunologically-induced non-specific reactivity due to:
   a presence of antibody that reacts with either the same (shared reactivity), or similar (cross-reactivity), epitopes on different antigens, or
   b presence of an antibody which reacts with unrelated antigen which is coincidentally present in the assay system (eg if impure antigens are used, or if antibodies to mice, sheep, or goat immunoglobulins are present and reagents derived from these species are used as part of the assay).
2 Specificity problems associated with the nature of the clinical material being examined (eg specimens containing mucus, pus or microbial contaminants may adsorb directly to a solid phase and bind non-specifically any detector protein or conjugate used).
3 Biological components present in clinical material which may affect the specificity of the test reaction (eg rheumatoid factor, natural haemolysins, B_2 microglobulin, serum lipids). Where such components are known to influence an assay they should be sought, and removed or inactivated if found. Suitable test controls

Immunoassays

should be included in the assay to ensure that inactivation or removal has been achieved.

4 Adverse effects due to treatment of specimens prior to testing (eg heat inactivation, addition of sodium azide, recalcification of plasma).

Where an assay has been properly evaluated, most of these potential problems will have been investigated and documented. However, in practice, the recognition of such problems and the incorporation of suitable controls largely relies on the expertise and clinical awareness of the laboratory staff performing the assay.

Confirmation of test results

As discussed above, the specificity of a test may be influenced by factors for which it is not always possible to include appropriate controls. It is therefore important that any unexpected positive or negative result, or result which will have a profound clinical or social impact, should be confirmed. Methods by which test results can be confirmed include:

1 Repeat testing of specimens not initially tested in replicate (this is not very satisfactory, as the error may be reproducible).

2 Use of a different assay for the same analyte (eg confirmation of a microagglutination result by immunofluorescence in *Legionella* serology).

3 Use of an assay for a different marker of the same infection (eg demonstration of anti-hepatitis B core IgM to confirm the diagnosis indicated by a HB_sAg positive result).

4 Confirmation of positive reactions by blocking or neutralization tests (eg the verification of anti-HIV screen positive results using specific blocking reagents).[19]

5 Purification of the analyte and repeat of the assay (eg sucrose density gradient fractionation of serum to confirm the presence of anti-rubella IgM).

6 Referral to reference laboratory for confirmation.

Reference laboratories

Although a test may be reproducible within a laboratory, the results obtained may not agree with those of another laboratory. This inter-laboratory variation is largely overcome by the use of reference material (see page 101). However, a second approach is periodically

Quality control in the microbiology laboratory

to send specimens to a reference laboratory. While not necessarily the final arbiters, reference laboratories should be able to offer a high degree of consistency against which a laboratory can compare its results. Also, as reference laboratories examine specimens from a wide range of sources, they may become aware of any developing problems, such as the declining quality of a particular reagent, earlier than would the referring laboratory.

External quality assessment

Participation in external quality assessment exercises can provide a laboratory with additional information about the performance of kits employed, and the quality of service offered. These exercises involve a simple exchange of material between laboratories, or regular participation in a regional or national quality assessment scheme.

The UKNEQAS for Medical Microbiology (see Chapter 1) distributes clinical material for a wide range of diagnostic procedures including immunoassays. Participation in this type of scheme provides an opportunity for a laboratory to monitor the performance of internal quality control practices, to assess the level of technical expertise in the laboratory, and to assess the performance and reliability of test systems involved. Comparison of results obtained from different laboratories may reveal errors overlooked in internal quality control programmes, or may indicate inadequacies in the assays used. Problems that have been identified with serology distributions include:

1 incorrect calibration of working standards
2 inadequate operational controls
3 inappropriate operational controls
4 no statistical analysis of quantitative assay results
5 poor standardization of immunoassays.

Information collated about the tests employed can reveal trends in the range of tests employed and the pattern of testing, which may be conducive to improved standardization.

Maximum benefit can only be obtained from quality assessment exercises if quality assessment specimens are processed in the same manner as routine clinical specimens. Failure to do so could result in the failure to recognize problems affecting the performance of the immunoassays. Similarly, quality assessment specimens should not be used as a substitute for an internal quality control programme, which should be implemented on a daily basis.

If it is not possible to assess the performance of a particular immunoassay or diagnostic procedure because quality assessment specimens are not readily or regularly available, then clinical material should be exchanged with another laboratory and the results obtained in each compared.

Conclusions

The time and effort required to perform quality control is often given grudgingly. A carefully-implemented scheme will in the long term, however, ensure that the laboratory's work is of a high standard, is cost-effective and will improve patient management. In this chapter we have attempted to provide broad guidelines for such a QC scheme for immunoassays. The exact QC programme used will, of course, have to be tailored to suit the prevailing clinical and laboratory requirements, but the salient features will remain the same.

Finally, it should be remembered that QC programmes are not static and should be regularly reviewed and updated to take account of the increasing range, complexity and improved performance of the newly-emerging immunoassays.

References

1   Grange JM, Fox A, Morgan NL. Immunological techniques in micro-biology. Society for Applied Bacteriology Technical Series 24. Oxford: Blackwell Scientific, 1987.
2   Voller A, Bartlett A, Bidwell D (eds). Immunoassays for the 80's. Lancaster: MTP Press, 1981.
3   Collins WP. Alternative immunoassays. Chichester: Wiley, 1985.
4   Consumer Protection Act 1987. Chapter 43; Part 1. London: HMSO, 1987.
5   Brian JA. The serologic diagnosis of viral infection. *Arch Pathol Lab Med* 1987; **111**: 1015–23.
6   Sever JL. Automated systems in viral diagnosis. In Bachmann PA (ed.) New developments in diagnostic virology. *Current Topics in Microbiology and Immunology* 1983; **104**: 57–75.
7   Greiff D, Rightsel WA. An accelerated storage test for predicting the stability of measles virus dried by sublimation *in vacuo. J Immunol* 1965; **94**: 395–400.

Quality control in the microbiology laboratory

8   World Health Organization Expert Committee on Biological Standardization. 32nd report. *WHO Technical Report Series* 1982; **673**.
9   Strike PW. Medical laboratory statistics. Bristol: Wright, 1981.
10 Colvin HM, Ghysels G, Leblanc A. Guidelines for evaluation and establishment of reference methods in microbiology and immunology. World Health Organization, Lab/82.3, 1982.
11 Anon. HIV(LAV/HTLV III – 'Aids virus') antibody in diagnostic reagents and quality control and calibration materials. Health notice (HN(86)2S), Health Services Division. London: Department of Health and Social Security, 1986.
12 Finney DJ. Statistical method in biological assay. London: Griffin, 1978.
13 Pattison JR (ed.). Laboratory investigation of rubella. Public Health Laboratory Service Monograph Series 16. London: HMSO, 1982.
14 World Health Organization. Biological substances: international standards and reference reagents 1986. Geneva: World Health Organization, 1987.
15 National Institute for Biological Standards and Control. Biological reference materials 1986–87. Potters Bar: NIBSC, 1986.
16 Chard T. An introduction to radioimmunoassay and related techniques. Laboratory techniques in biochemistry and molecular biology 6. Oxford: Elsevier, 1987.
17 Ekins RP. Quality control and assay design. In: Radioimmunoassay and related procedures in medicine 2. Vienna: International Atomic Energy Agency, 1978, 39–54.
18 Rodbard D. Statistical quality control and routine data processing for radioimmunoassays and immunoradiometric assays. *Clin Chem* 1974; **20**: 1255–70.
19 Hellings JA, Theunissen H, Keur W, Siebelink-Liauw A. New developments in ELISA verification of anti-HIV screening of blood donors. *J Virol Meth* 1987; **17**: 11–17.

9 Virus isolation

JM Darville and EO Caul

Introduction

Virus isolation is the detection by replication and therefore amplification in living cells of infectious virus from a clinical specimen. In some instances it is the most direct and definitive method of virus diagnosis. When positive, it is likely to be significant in terms of the patient's illness except perhaps in the case of persistent viruses and sometimes enteroviruses in children. Although usually more rapid than conventional virus serology, virus isolation is rarely rapid enough to influence the management of the illness and so, with the development of antiviral agents, more emphasis is being placed on viral antigen detection by immunofluorescence and enzyme linked immunoassays (ELISA) (and to a lesser extent on nucleic acid detection) in clinical samples. Nevertheless, these methods are not available for all viruses and, also, it is essential in some instances to isolate a virus (eg influenza) in order to monitor epidemiological trends. Furthermore, isolation is still the 'gold standard' against which the newer technologies are measured: it would be unwise to abandon the means to evaluate commercial products continually.

A good case can be made, therefore, for continuing to isolate viruses even though this may not contribute much to patient management. If done, it must be done well, requiring attention to quality control in the areas considered in detail below. The aim of this control is to maximize virus isolation rates using optimum techniques.

Quality control in the microbiology laboratory

General working practices

Although technically easy to perform, successful cell culture and virus isolation require considerable knowledge and experience on the part of the operator, who should be closely supervised by at least a senior MLSO or equivalent during training. To a considerable extent, the skills required are intuitive – much like the green fingers of gardeners. Nevertheless, an eye for cytopathic effects can be developed, ideally under the tuition of an experienced senior colleague. Furthermore, good results are encouraged by the formation of and adherence to straightforward laboratory procedures. In particular, scrupulous attention should be paid to the use of aseptic techniques, and basic training in microbiology is necessary. If accidental contamination of materials, media or cultures occurs or is suspected, then there should be no compunction about discarding them. Flaming of glassware, although traditionally established, is useless since it generates insufficient heat for sterilization. Safety cabinets should be kept uncluttered, clean, and the working surface washed down with disinfectant (eg 2% Stericol®) at least daily, and whenever spillage occurs. They should be fumigated periodically and monitored weekly for satisfactory air-flow. For the selection of appropriate safety cabinets, the reader is referred to the Health Services Advisory Committee's *Safe working and the prevention of infection in clinical laboratories* (see Further Reading). Again, when only culture protection is required, rubber gloves are not necessary. It is important, however, to ensure that hands are regularly washed, especially after soilage or usage in contaminated areas. Long hair and beards should be restrained both to protect cultures and to prevent accidents from machinery or bunsen burners. Finally, the success of virus isolation is promoted by thorough preparation. This should include monitoring the cleanliness, sterility and absence of toxicity of all materials and media and confirming that all reagents and cell cultures function as they should before use.

Materials and media

The first requirement is to ensure that materials and media are satisfactory before use in cell culture or virus collection and transport.

Glass and plasticware

There is now a strong trend away from reusable glassware towards disposable plasticware. One obvious reason for this is laboratory safety. In addition, plasticware is more convenient to use and is cheaper than glassware. However, problems of waste disposal and the exhaustion of non-renewable resources may, in the future, shift the balance back towards glass.

Generally, the sterility, non-toxicity and quality of plasticware to support cell growth are guaranteed by the manufacturers. Nevertheless, it is advisable to test new batches and especially new suppliers for these properties by culture through at least three passages of the more fastidious cell types, eg human fibroblasts, Vero cells, in the vessels in parallel with proven stock. Much more care is needed with reusable glassware (including caps, seals and bungs): wash new stock thoroughly to remove toxic coatings, and sterilize and wash stock in use to remove material from its previous use. Washing procedures must be monitored for efficiency, and rinsing must thoroughly remove residual detergent. Before reuse, re-sterilize the glassware, monitoring the procedure using autoclave tape or Browne's tubes etc.

Instruments

In the preparation of primary cultures and cell strains, all instruments must be cleaned and sterilized before use.

Water supply

Good cell culture is dependent on deionized or glass-distilled water, and some fastidious cells require Analar-grade water. For most routine laboratory cell cultures, deionized water is entirely satisfactory. Check deionizer function by monitoring conductivity as described by the manufacturer and sterilize the water. Periodic checks on pH are advisable. In some areas deionization is not enough and water double-distilled in glass stills is required.

Culture medium

Cell culture medium is now usually purchased from manufacturers and is rarely prepared in laboratories. If prepared as recommended problems are rare, but some products need careful adjustment of pH.

Quality control in the microbiology laboratory

All new batches of medium should be tested for their growth-promoting properties by colony-counting; problems may arise from lack of essential ingredients or from sub-lethal toxic substances.

Calf serum – fetal, newborn

This is a perennial source of problems in cell culture. Serum may contain viruses, eg mucosal disease virus, or cytotoxic factors, non-specific antiviral factors or antiviral antibodies. New batches should be tested for appropriate viruses by immunofluorescent staining and for sterility (including absence of mycoplasma) before purchase is complete. Heat treatment may inactivate toxic factors. Subculture cell lines for a minimum of six passages to ensure that serum used for cell growth is not cytotoxic. This should be controlled with a known non-cytotoxic serum.

Antibiotics

Generally, these provide no problem. However, they must be prepared and used as recommended by the manufacturer. At some concentrations, antibiotics are toxic for mammalian cells. Although penicillin is not clinically effective against chlamydia, it does inhibit their growth in cell culture and so should not be incorporated in the medium of cells used for isolating these agents.

Other additives

Amino acids and vitamins may be obtained sterilized, or filter sterilized in the laboratory. Bicarbonate buffer may be obtained commercially or, more commonly, prepared in the laboratory. As it is steam-sterilized, take particular care to confirm sterility. If toxicity problems persist when all other reagents have been cleared, then test the phenol red. Similar testing should be applied to non-volatile buffers such as HEPES.

Washing, digesting agents

Phosphate-buffered saline is now universally available in tablet form. All that is required is to make it up correctly and ensure sterility.

Similarly, ethylene diamine tetra-acetic acid (EDTA), trypsin and collagenase are available from manufacturers for dispersal of cells for culture purposes. It is necessary to ensure that trypsin is free from animal rotaviruses and so, in any attempts to culture rotaviruses, this should be checked by examining uninoculated cultures using immunofluorescent staining. Furthermore, such a virus in cell cultures may potentially interfere with the isolation of human viruses.

Sterility checks

After sterilization, check all batches of reagents used in cell culture for sterility before use by standard bacteriological techniques, inoculating both solid and liquid media with 5–10ml of reagents and incubating at 22°C and 37°C for five days. Only when all materials and media have been shown to be satisfactory should they be used for virus isolation.

Virus isolation systems

Cell culture

Primary/secondary cultures   Of these, only human embryo kidney cells and human embryo fibroblasts are commonly produced within virus laboratories, and their availability is often limited by a low or irregular flow of fetal material. Because human embryo kidney cells are essentially unpassageable, their preparation incurs a high workload. Furthermore, there may be considerable variation in sensitivity from batch to batch, which must be monitored (see page 119). They must, of course, be prepared and handled aseptically.

Animal cells of this type are rarely produced in house because of the lack of availability of animals and suitable handling facilities, let alone appropriately trained and licensed personnel. These cells are usually supplied from commercial sources who may maintain colonies of animals for this purpose. Such cells may be more uniform than those from human sources, but nevertheless sensitivity (and sterility) should be tested before use, or at least be monitored during use. Endogenous viruses in primary cells may lead to clinically false positive results in virus isolation, or to interference. This too should be monitored. Primary monkey kidney cells may be infected with polyomavirus (SV40), myxovirus (SV5) or herpesviruses; if these are present the cultures should be discarded. Before using these cells for isolating respiratory viruses, test representative cultures for endogenous haemabsorbing activity caused by myxoviruses.

Quality control in the microbiology laboratory

Cell strains – semicontinuous These may be obtained from suppliers but, since they are capable of extended passage, they are often produced in house from human fetal material. As before, aseptic precautions are essential in their production. The possibility that these cells, too, may contain endogenous viruses should not be overlooked. Furthermore, it must not be forgotten that the fetal material may be contaminated with HBV, HIV or other human pathogens, so the use of gloves at this stage in their preparation is advisable.

Since these cells are passaged several times before senescence, the chance of contamination with fungi, bacteria or mycoplasmas is greater than with primary cultures. The operator is a common source of mycoplasmas, and further protection may be afforded by the use of masks, at least during preparation of the primary cells. Check these regularly for such evidence of infection as turbidity, visible colonies, organisms visible by microscopy, altered cell growth characteristics and pH changes. Some low-grade contaminants may only become obvious on antibiotic-free medium or by subculture of medium or cells in or on fungal or bacterial growth media. Therefore stocks must be checked periodically for low-level contamination with standard bacteriological techniques, since the biggest problems are caused by insidious, slow-growing organisms.

As with primary cells, new batches should be tested before use for sensitivity to a range of viruses known to replicate in these cells.

Cell lines These immortal cells may be established in house from the previous cell types, but in practice most laboratories use well known lines such as Hela, HEp-2 and Vero cells. They can be obtained from commercial suppliers but are often, if not usually, obtained as gifts from other laboratories or kept in liquid nitrogen storage (at 'low passage') within the laboratory.

Cell lines may be less sensitive and show broader specificity than do primary cells or cell strains. However, they are generally convenient and give reproducible results. Nevertheless, clonal variation can develop on repeated passage and so cell lines too should be tested for their sensitivity before use. In particular, it is essential to evaluate batches of HEp-2 or Hela cells before the RSV season each year. This may be done by inoculating ten-fold dilutions of a reference strain into at least three replicate tubes each and measuring the end point.

Perhaps the main problem with cell lines is that their immortality

Virus isolation

magnifies the chance of their becoming persistently contaminated, especially as they may be passed from one laboratory to another. Contamination may even lead to increased sensitivity to some viruses (perhaps by inhibiting interferon synthesis), but it is not advisable to rely on such fortuitous enhancement, nor to risk the spread of contamination to less forgiving cells. Periodically, therefore, test stocks of continuous cells for contamination with mycoplasma by staining coverslip cultures with Hoechst or quinacrine and examine for fluorescent staining. If positive, retrieve fresh stocks from storage or purchase (since they are relatively cheap) from suppliers.

General points   All cell stocks must be regularly monitored and tested for microbial contamination. Destroy any found to be contaminated: it is not worth while attempting to 'cure' cells. Possible exceptions to this rule are mycoplasmas, which do not produce gross effects but whose presence may be inferred by more rapid cell growth and reduction in pH. This may be confirmed by staining, culture or enzyme detection. It is sometimes possible to eliminate mycoplasma by treating with kanamycin for one month.

Ideally, different cell types should be passaged on different days in order to reduce the chance of contamination of one cell type with another. Failing this, then at least process primary cells before strains, and strains before lines. Furthermore, each bottle of culture medium should never be used for more than one cell type. Another guard against problems caused by contamination is to maintain each cell type as two separate but parallel stocks, each split and changed on different days. Check the identity of permanent cell lines by examining the karyotype and by testing for glucose-6-phosphate dehydrogenase, present in Hela cells but not other cell lines.

All cells should be passaged and fed at regular intervals. They should be passaged at neither too high nor too low a ratio nor allowed to become overgrown (primary cells and cell strains especially). This is to ensure minimal disturbance of the cells' growth and metabolism.

Cell storage

Many laboratories maintain in liquid nitrogen stocks of low passage uncontaminated cells from which routine cultures may be replenished. These cells should be frozen at 10^6/ml in medium containing

Quality control in the microbiology laboratory

dimethylsulfoxide (DMSO) (or other similar agent to prevent damage on freezing) at 1°C per minute. For safety reasons, store in plastic vials in vapour phase refrigerators. In contrast, thawing should be rapid, eg by immersion in water at 37°C. Before opening, swab vials with ethanol and then transfer their contents aseptically to fresh growth medium in culture flasks. Once the cells have adhered to the flask, change the medium to reduce the chance of toxic effects of the DMSO etc on the growing cells. These procedures ensure maximum cell survival. It is essential to maintain comprehensive records on stored cells: ie, passage number, date of storage and the results of recovery.

Animals and eggs

The use of animals and embryonated eggs for routine virus isolation continues to decline. When they are used (eg for Coxsackie virus A isolation), it should be established that they are uniform, pathogen-free and able to support the growth of the virus types sought. Perhaps the most important point to note is that the use of eggs over 10–11 days old now requires a Home Office licence.

Collection, storage and transport of specimens

The aim of control here is to maximize the chance of a virus present in the patient still being infectious when inoculated into the cell cultures. The importance of this to the virus isolation rate should not be underestimated, although too often it probably is.

Transport medium (TM)

This may be liquid or semi-solid; it must be isotonic. The inclusion of protein stabilizes virus particles and may (in the case of milk saline at least) help to preserve cells in culture. However, take care to ensure that this protein is neither cytotoxic nor inhibitory to viruses. Antibiotics are optional. Appropriate controls on TM in addition to those mentioned above are for sterility and for the ability of a range of viruses to survive in it.

Swabs, containers and kits

Problems rarely arise with sterilized cotton wool and wood or wire swabs. However, alginate swabs inactivate herpes simplex viruses and wooden sticks may interfere with chlamydia ELISA tests. Specimens of faeces, urine, sputum, aspirates and CSF are collected in containers sterilized either by suppliers or in house. For the users' convenience, the laboratory may supply complete specimen collection kits which, in turn, should improve the quality of specimens. The laboratory should seek to ensure that there is an efficient system for keeping users supplied and that old stocks of TM are discarded.

Education

Specimen quality can also be improved by ensuring that those responsible for collecting or ordering the collection of specimens know which specimens to collect in a given clinical condition and that isolation in general is unlikely to succeed more than four or five days after the onset of acute infections. Under ideal circumstances, specimens would be collected by laboratory staff; others should receive precise guidance on procedures.

Transport and storage

It is important that a 'cool-chain' from patient to laboratory can be maintained to avoid either overheating or freezing of specimens; this is particularly important for specimens suspected of containing labile viruses.

The ideal temperature is 4°C, although ambient temperatures will suffice for most viruses if transit times are short (except in hot weather). This applies when the specimen is in the ward or clinic, in transit or at the laboratory. Ideally, specimens should be inoculated on receipt, but since it is more convenient to batch them, they should be refrigerated until processed. Since transport is often unreliable, urgent or precious specimens should be collected by laboratory staff or sent by courier or taxi.

Only when materials, media, cell cultures, specimen collection and transport have been optimized should virus isolation be commenced.

Virus isolation

Batch testing

Before each batch of cell cultures is used and before the beginning of the season in the case of seasonal viruses, it should be established that each batch can support the growth of the expected viruses, at the

Monitoring in use

Each cell culture (ie flask, tube, well) should be examined microscopically and any unsatisfactory culture discarded. The correct choice of cell types is vital to the success of virus isolation.

Inoculation

The specimen should be inoculated into a range of cell cultures, using two tubes of each cell type, able to support all the (known) viruses likely to be in it. This, of course, requires a full clinical history of the patient as well as information on the duration of the illness. Uninoculated controls may be included with each day's batch of cultures. Improve sensitivity by removing the maintenance medium and applying the specimen directly to the culture. This does, however, also increase the risk of toxic effects, and it is advisable to remove early (ie after one hour) the inoculum of specimens known or likely to be toxic, such as tissue extracts, faeces and urines. After inoculation, specimens should be stored frozen in case attempts at re-isolation become necessary. Those likely to contain labile viruses (herpes viruses, respiratory viruses) must be snap-frozen and stored at −70°C (or lower).

Incubation

Cultures are generally incubated static at 36–37°C. However, incubate representative cultures from respiratory specimens at 33°C and roll to improve the isolation of rhinoviruses particularly. Check incubator temperatures daily; preferably use a temperature recording system (correctly calibrated). Unless cultures become contaminated or aged (which with good quality control they should not), they should be incubated for as long as necessary to isolate slow-growing viruses (eg VZV, CMV, polyomavirus) from appropriate specimens.

Examination

Cultures should be examined regularly (ideally every one or two days) for the presence of CPE, toxicity, contamination and ageing: it

Virus isolation

is important to be able to differentiate viral effects from the others and to note that toxicity due to *Clostridum difficile* toxin in faecal specimens may be clinically significant. To improve the chance of recognizing CPE early, procedures for examining the whole monolayer in a methodical manner should be employed.

In order to prevent cross-infection or contamination, remove all positive and contaminated cultures before the negative cultures are changed. Positive cultures should be further processed in a separate cabinet or room if available. It is not worth attempting to save contaminated cultures unless the specimen is precious. For specimens which are intractably toxic, or in which viruses with late or slow CPE are suspected, isolation rates can be improved by blind passage. For viruses which remain strongly cell-associated (eg SSPE strains of measles), co-cultivation of whole tissue or cell suspensions with susceptible indicator cells should be attempted.

Keep complete records of inoculated cultures. This should include information on medium changes, passage and observation of cytopathic effect. Such records permit the detection and rapid rectification of problems.

Dealing with contamination

Clinical specimens not infrequently contain antibiotic resistant microorganisms which contaminate the cell cultures into which they are inoculated. Should this occur, then specimens can be treated with an appropriate antibiotic and/or passed through 200nm filters before attempting re-isolation. If this fails then further efforts are not justified. The final report should indicate the presence of antibiotic-resistant organisms.

A far greater problem is posed when whole batches of cells become contaminated. If a single cell type is involved, all these cultures must be discarded, followed by the introduction of fresh batches of cells grown in fresh medium. However, when more than one cell type is involved, either poor technique or, more likely, contaminated medium components will be responsible. In this case it is essential that virus isolation is suspended and not resumed until the source of contamination is investigated microbiologically and eliminated. This will inevitably lead to delays and backlogs, but if not instituted is likely to lead to problems of chronic contamination.

Virus identification

Although not strictly within the remit of this chapter, virus identification is the logical conclusion of virus isolation. Viruses may be identified by neutralization, immunofluorescence, inhibition of haemagglutination and haemabsorption and less frequently by the analysis of proteins and nucleic acids. Success here too requires strict observation of control procedures.

Testing the system

Although virus isolation may appear to be working well, there must be procedures for checking that this is indeed so. Firstly, there should be in-house disguised testing, conducted by a senior member of the laboratory staff. This should consist of the preparation and presentation of positive material as authentic clinical specimens, which may require the cooperation of the laboratory's customers. It is vital, however, that this is done with the full approval of virus isolation personnel, who should always be informed of results and encouraged to discuss them. Secondly, the laboratory should belong to an external quality assessment scheme. It is inevitable that specimens provided in such schemes receive greater attention, albeit unconsciously, than routine specimens. However, if attempts are made to minimize their preferential treatment, they serve as a measure of the overall standard of the laboratory and of how it compares with other laboratories.

Summary

In order to ensure the quality of the virus isolation service, straightforward procedures should be devised to minimize the chance of problems arising and to enable any problems which do arise to be solved quickly. These procedures must be adhered to and not modified without due consideration. The quality should be controlled both by examination of coded specimens from time to time and by participation in an external, overt scheme with full discussion of results.

Acknowledgements

We thank Dr APCH Roome, Miss MJ Usher and Mr JH Moule for their critical appraisal of the manuscript.

Virus isolation

Further reading

1 French LVM, Leland DS. Concepts of clinical diagnostic virology. In: Lennette EH (ed.) Laboratory diagnosis of viral infections. New York: Marcel Dekker, 1985, 1–39.

2 Grist NR, Bell EJ, Follett EAC, Urquhart, GED. Diagnostic methods in clinical virology, 3rd edn. Oxford: Blackwell, 1979.

3 Health Services Advisory Committee. Safe working and the prevention of infection in clinical laboratories. London: HMSO. [In press]

4 Landry ML, Hsiung GD. Primary isolation of viruses. In: Specter S, Lancz GJ (eds) Clinical virology manual. New York: Elsevier, 1986, 31–51.

5 Lee IC. Quality control in clinical virology. In: Specter S, Lancz GJ (eds) Clinical virology manual. New York: Elsevier, 1986, 3–14.

6 Malherbe HH. Role of tissue culture systems. In: Lennette EH (ed.) Laboratory diagnosis of viral infections. New York: Marcel Dekker, 1985, 41–3.

7 Schmidt NJ. Cell culture techniques for diagnostic virology. In: Lennette EH, Schmidt NJ (eds) Diagnostic procedures for viral, rickettsial and chlamydial infections, 5th edn. Washington: American Public Health Association, 1979, 65–139.

8 Smith TF. Specimen requirements: Selection, collection, transport and processing. In: Specter S, Lancz GJ (eds) Clinical virology manual. New York: Elsevier, 1986, 15–29.

9 Versteeg J. A colour atlas of virology. London: Wolfe Medical, 1985.

10 Electron microscopy

A Curry and DJ Wood

Electron microscopy (EM) is the gold standard for the detection of faecal viruses because it is the only 'catch all' method available. Its importance in this area and in the rapid identification of viruses in skin lesions, particularly from the immunocompromised, means that it will continue to play an important diagnostic role for some years to come. Alternative diagnostic methods, eg ELISA, have been compared with electron microscopy for the detection of group A rotaviruses and enteric adenoviruses, which are arguably the most important causes of viral gastroenteritis in children: these new techniques have been shown to be at least as sensitive. However, group B[1] and group C[2] rotaviruses, which cannot yet be detected by commercial kits, can be visualized by electron microscopy. Any new commercial assays should be compared with electron microscopy before they are used routinely. In addition, although electron microscopy is not an ideal method for the detection of the important Norwalk-like group of viruses, it is the only diagnostic method available to routine laboratories.

In the field of diagnostic virology, EM is primarily restricted to the examination of negatively stained material under the transmission electron microscope. To achieve the high standards necessary, electron microscopy requires a thorough understanding of the operation of the microscope, preparative techniques, photography, virus morphology and the clinical effects of virus infection. For these reasons, the EM laboratory should be managed by a senior member of staff who will also be responsible for implementing quality control (QC) procedures. The procedures needed to maintain the quality of EM tests can be conveniently considered under several headings.

Collection of specimens

Liaison with clinicians or environmental health officers on the collection of specimens should be part of the QC procedure. Viruses are usually most plentiful early in the course of the disease and certain gastroenteritis viruses, particularly the Norwalk-like virus (SRSV), are normally excreted for only about 48 hours after onset of symptoms. It is, therefore, vital to obtain specimens as quickly as possible so that there is a reasonable chance of detecting viruses by electron microscopy. With outbreaks of diarrhoea and/or vomiting involving large numbers of specimens, it may not be possible to examine all the specimens submitted by electron microscopy. It is, therefore, important that those collected within 48 hours from patients with symptoms are examined first.

Preparative methods and reagents

Specimen preparation

Preparation of faecal material should be by a method which will both partially purify and concentrate any virus particles, eg, the double spin centrifugation method[3] or the ammonium sulphate precipitation method.[4]

Negative stains

The quality of the negative stains used (usually phosphotungstic acid), can be judged from the quality of the staining when the grids are examined (see page 130). However, the effect of various stains and their pH on virus morphology may not be so obvious. It has recently been demonstrated that some rotaviruses can be morphologically damaged by certain stains and at certain pH.[5,6] These observations may have important implications for routine diagnostic use, therefore the pH of stains should be monitored regularly.

Immune EM

Solid phase immune electron microscopy (SPIEM) or immune clumping have shown the ability to both improve the detection of some enteric viruses (eg SRSVs) or to serotype others (eg group F adenoviruses). The characteristics of antisera for such immune EM tests must be well defined before use, particularly when utilizing human sera. Thus, for use in typing assays antisera must be free of reactivity against antigens that are serologically distinct but morphologically identical to those of the virus of interest. For example, antisera used to identify group F adenoviruses should have no reactivity, by immune EM, against a variety of non-group F adenoviruses.[7]

The use of control reagents is vital in immune EM. Ideally, the pre-immune serum from the animal used to raise the specific antisera should be used but if this is not available a serum from a different, unimmunized animal of the same species can be used. In either instance the control antiserum should be used at the same dilution as the specific antiserum. A control in which antiserum is replaced by buffer should also be included, especially in quantitative work, since control antisera can inhibit binding of viruses to grids[8] and the grids prepared with buffer only will more precisely represent the numbers of particles in the original sample. Criteria for the interpretation of results must be developed and strictly adhered to.

In addition to control antisera, known positive and negative standard antigens should be tested in parallel with each batch of unknown antigens to verify the test results.

Immune electron microscopy results should only be accepted if all controls are satisfactory. Furthermore, artefacts must be recognized and excluded, eg, in clumping immune EM techniques 'pseudo clumps' due to viruses lying close together by chance must be differentiated from true immune complex clumps.

Operation of the microscope

Electron microscopes need constant fine tuning to maintain performance. These adjustments should be within the capabilities of the EM staff. Since image quality sufficient to resolve characteristic features of virus morphology is essential, tests should be made regularly as described in detail below.

Checking column alignment

Misalignment of the column can affect image quality and this should be checked before attempting to correct astigmatism. With the instrument magnification control set to a high magnification, locate and place an image feature in the centre of the field of view. Move the medium focusing control (objective focus) over its complete range and observe image movement. The image should hardly move at all if column alignment is satisfactory. Column realignment can be a complicated and confusing procedure unless the operator is very familiar with the instrument. Alignment procedures are covered in the operators' manual but, if in doubt, leave this task to the service engineer.

Astigmatism

Astigmatism is a common fault that may obscure essential morphological detail. There are several causes of astigmatism, the most common of which is contamination which settles asymmetrically on the lens bore, objective aperture or specimen carrier. Other causes are specimen-induced astigmatism, an out of centre objective aperture and imaging occurring very near the periphery of the specimen grid. Curing astigmatism depends fundamentally on the operator recognizing the problem and then locating and correcting the cause. Lens bores, specimen carriers and apertures should be clean and contamination free. Specimen carriers and apertures (unless self-cleaning thin film apertures are used in the objective) should be cleaned on a regular basis in accordance with the manufacturers' cleaning instructions.

Astigmatism, if detected, must be corrected. Of the two methods in common use, the first utilizes a special 'holey carbon film' grid, whereas the second utilizes 'background granularity' present in the specimen or its support film. The latter is of use only on high magnification, high resolution instruments. The manufacturers' instructions for correction of astigmatism should be followed. Some operators have difficulties with these procedures, so supervision of correction and regular practice are needed to achieve competence.

Drift

Image drift can be caused by many faults, including excessive movement of the EM specimen stage, movement of the grid in the specimen holder, poor adhesion of the support film to the grid bars, an adjacent hole in the support film, heating effects of the beam and overlong photographic exposure. Carbon coating of the plastic support film dramatically reduces thermal stretching but this can produce a hydrophobic surface (ie material adheres poorly). When present to only a slight degree, drift can produce an effect mimicking astigmatism in the micrograph. Operators must be aware of drift and attempt to eliminate the causes.

Focusing

Precise focusing of the image on an electron microscope screen is essential but is not easy and requires a great deal of practice. Slight underfocus of the image gives maximum contrast, the most desirable setting. Focusing must be performed quickly, as build-up of contamination or beam etching can obscure the detail one is trying to record. If focusing takes the operator some time to accomplish then the anticontaminator ought to be used. With practice, focusing becomes quickly attainable.

Magnification calibration

Sizing of virus particles is an important consideration in terms of accurate diagnosis. All electron microscopes have marked magnification steps. These, however, should not be regarded as true values because of the phenomenon of lens hysteresis. Providing each step is stable and regularly recalibrated, sizing can be accomplished with confidence. Most instruments have a 'normalization' button which should be used if accurate measurement is required. This control increases the current in the imaging lenses to saturation and then returns the current to the desired instrument magnification value to achieve a reproducible figure.

Grid distortion and position in the lens can also cause magnification variation. Some instruments have an adjustment for specimen height which must be set to the eucentric position to achieve reproducible results.

Many calibration specimens are available, but only a few are useful for the magnification range used in virology. The ruled diffraction grating (line grating with 2160 lines/mm) is useful for the calibration of magnification values up to 20K but not higher ones. For magnifications above 25K the beef catalase calibration specimen is more appropriate as it has crystalline spacings (6.85nm and 8.75nm ± 0.5%) which are of the same order of magnitude as the resolution necessary to distinguish detail in virus particles.

Several photographs should be taken of these calibration speci-

Electron microscopy

mens which should then be developed and the negatives selected to show the best lattice lines. These should be measured using a loupe with a graticule divided into 0.1mm divisions. The following formulae should be used to calculate the actual magnification value.

$$\text{Actual magnification} = \frac{\text{average spacing (mm)} \times 10^6}{\text{spacing distance (nm)}}$$

Once the instrument is calibrated, the size of a virus (or any other object) can be calculated with the following equation.

$$\text{Size (nm)} = \frac{\text{size on negative (mm)} \times 10^6}{\text{calibrated magnification}}$$

Detection and recognition of virus particles

The quality of results in electron microscopy is fundamentally dependent on the observational skills of the electron microscopist. A good observational strategy when examining a grid is necessary to provide assurance that a virus present in sufficient concentration will be detected. For example, each grid should be scanned initially at a low magnification to select an appropriate area for examination at high magnification (Figure 10.1). The presence of cellular or bacterial debris, which is found in most clinical specimens, is a good indicator that the specimen has adhered to the grid. However, the presence of localized pools of negative stain, which are intensely electron dense (Figure 10.2), indicates poor spreading of the negative stain and could obscure viruses present in a sample. Therefore the absence of cellular/ bacterial debris and/or poor staining are indications that the preparation should be repeated. The magnification used to search for viruses is a compromise between a magnification high enough to allow virus particles to be identified and low enough to allow a sufficient area of the grid to be scanned in a given time period. This magnification is a matter for personal preference but should be in the range of 30,000– 60,000× on the screen. It is more important that one particular magnification is used routinely and that the operators learn to recognize viruses at that magnification. It is also essential that the microscopist is taught to search the grids in a systematic fashion so that new areas are always being examined. Many viruses are readily observed because of their geometric construction and ability to stand out from surrounding cell or bacterial debris (Figure 10.3). However,

Quality control in the microbiology laboratory

Figure 10.1 Low power electron micrograph of a negatively stained grid showing areas too densely stained (bottom left) and well stained areas which are more likely to reveal viruses.

Electron microscopy

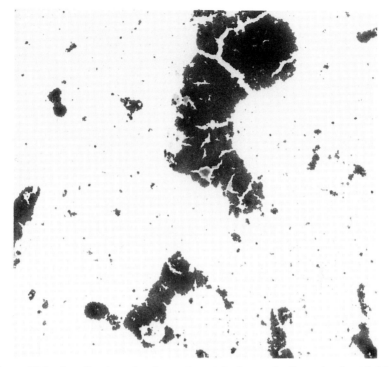

Figure 10.2 Localized pools of negative stain from a badly stained grid. No virus particles were seen on this grid. However, rotavirus and astrovirus were both readily observed in a repeat, well stained grid. Magnification ×150,000.

the electron microscopist must have an appreciation of the full range of virus morphology, which can be taught from standard texts, a collection of photographs or, more effectively, on the microscope screen from grids made from stored positive specimens. Stored positive grids can also be useful but these do deteriorate with time.

Another good working practice is to photograph the first virus or virus-like particle observed, since the EM negative is the only permanent record of that observation. Also, finer detail is present in an EM negative than in the screen image, thus use of the microscope only as a sophisticated virus viewing device is to be discouraged. At least two exposures of different virions or, preferably, groups of virions should be made for each morphologically distinct virus detected. However, when small round viruses are detected, more exposures should be

Quality control in the microbiology laboratory

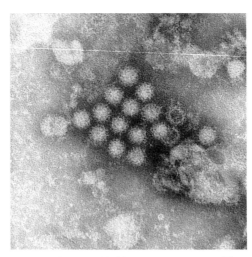

Figure 10.3 A group of Norwalk-like viruses (SRSV). These ragged-edged viruses readily stand out from the background. Magnification ×150,000.

made since the characteristic features of some small round viruses (eg astroviruses, caliciviruses) are only seen on a minority of particles (Figures 10.4 and 10.5). An interim classification of small round human faecal viruses has been proposed by Caul and Appleton[9] and their work should be consulted if the characteristics of the various small round viruses are not understood. Operators should not move on to the next specimen once a virus has been found. Many specimens may have several different viruses present.

It is important to impose a time limit on the high magnification scan after which, if no virus particles are seen, the specimen is reported as negative, as extended examination is unlikely to be productive. Ten to twenty minutes is a reasonable limit for faecal specimens although skin lesions may require longer.

The presence of some viruses can be reported on the basis of observations made directly from the microscope screen but it is essential that photographic negatives of all viruses should be assessed by an experienced electron microscopist. The diagnosis should be confirmed only if the particle shows the correct morphology and size range and its presence is compatible with the source of sample and clinical details.

If an experienced electron microscopist is not available in the laboratory, prints of particles should be sent to reference laboratories

Electron microscopy

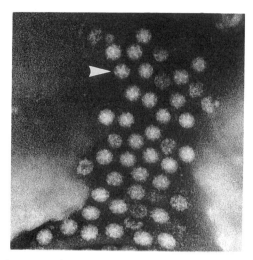

Figure 10.4 A group of astroviruses showing that not all particles feature the classic morphology. Arrowed particle shows characteristic surface star. Magnification ×150,000.

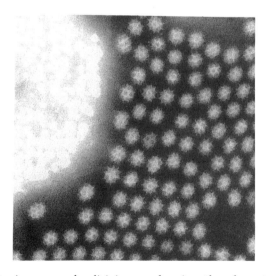

Figure 10.5 A group of caliciviruses showing the characteristic surface hollows, which fill with negative stain. Magnification ×150,000.

Quality control in the microbiology laboratory

for confirmation until sufficient experience has been developed locally. The electron microscopist must also, therefore, develop good darkroom working practices to produce good quality prints (see below).

A result of 'no virus seen' is more difficult to check but this should be done at least during initial training of microscopists. Review of grids by an experienced electron microscopist is the most effective method which should be continued until the microscopist is confident of the quality of results.

At some times of the year there may be a dearth of positive material with a potential concomitant loss of confidence in finding viruses. This can be overcome by occasional examination of known positive samples by routine procedures. Positive samples should be stored frozen unless they are known to contain Norwalk-like (SRSV) or calicivirus, where freezing often destroys the virus morphology. With such samples, storage at +4°C with a layer of liquid paraffin on top to inhibit fungal growth is recommended.

The quality control procedures outlined above can also be checked by participation in the periodic PHLS external quality assessment scheme in which specimens of known, but undisclosed, content are introduced into the laboratory and processed in the routine way. Failure in such schemes often indicates failings in internal quality control procedures.

Photography

Film exposed in the electron microscope needs to be developed and perhaps printed. Quality control must also be exercised here, so that the negatives and prints are of acceptable quality. It is important in photography that temperature, dilution, timing and length of storage times of reagents are all in keeping with the manufacturers' recommendations or are based on experience.

To achieve good photographic results, the image on the EM screen must be in focus, evenly illuminated, free of astigmatism and drift, and of sufficient contrast to show the virus morphology. Once an exposure is made, details of specimen number, patient name, virus or viruses found and instrument magnification should be recorded in a book or sheet that is kept with the plates or film. The negatives produced from the microscope should be well exposed without being either under- or over-exposed (ie almost transparent or very dense) as

Electron microscopy

this can lead to difficulties at the printing stage. A series of test plates should be taken (usually soon after microscope installation but repeated periodically if density is found to vary) with different instrument emulsion sensitivity settings in order to find the best level of exposure. A high contrast developer with good keeping properties is recommended (eg Kodak D19) for film development. After development, it is particularly important to see that the negatives correspond to the details on the record sheet. Negatives should be protected in suitable bags and labelled. They should be stored in dry, well ventilated conditions and, when handled, touched by the edges only. Not all negatives need be printed but good ones should be noted for future use.

If correctly exposed and developed, the photographic negative will show much fine detail and contain the range of contrast (tones) present in the specimen. With practice, the examination of such negatives alone can lead to definitive viral identification without printing. The surface star on astroviruses, for example, can often be more easily seen on the negative, whereas a poorly produced print can obscure this essential detail. Thus, printing of negatives must be skilfully performed, to ensure that the essential morphological detail present on the negative is reproduced on the print, with its more restricted range of tonal reproduction. Many micrographs from inexperienced or inadequately trained operators have too much or too little contrast. The examples of viruses shown in this chapter are regarded as being of acceptable photographic quality and are enlarged to the same overall magnification (150,000 ×) to make comparison easier (Figures 10.3, 10.4, 10.5)

Training

The quality of results in EM is critically dependent on the individual skills of the electron microscopist. The electron microscope laboratory should be run by a senior member of staff with a special interest in EM plus, ideally, a junior under tuition so that maximum continuity of service can be maintained. The appropriate skills need to be acquired by the development of good working practices and teaching of such procedures should be seen as an essential part of the quality control process. This requires a period of intensive one-to-one tuition by an experienced electron microscopist. Initial or advanced training can also be accomplished by sending designated staff to a larger laboratory for training.

Acknowledgements

We would like to acknowledge Dr EO Caul and Mrs Hilary Cotterill for their helpful criticisms. We would also like to thank Miss Sandra Corbishley and Miss Karen Griffiths for typing the manuscript.

References

1　Hung T, Chen GM, Wang CG, Chou ZY, Chao TX, Ye WW, Yao HL, Meng KH. Rotavirus-like agent in adult non bacterial diarrhoea in China. *Lancet* 1983; 2: 1078.
2　Caul EO, Ashley CR, Darville JM, Bridger JC. Group C rotavirus associated with fatal enteritis in a family outbreak. *J Med Virol* 1990; **30**: 201–5.
3　Riordan T, Craske J, Roberts JL, Curry A. Food borne infection by a Norwalk-like virus (small round structured virus). *J Clin Pathol* 1984; 37: 817–20.
4　Caul EO, Ashley CR, Eggleston S. An improved method for the routine identification of faecal viruses using ammonium sulphate precipitation. *FEMS Microbiol, Letters* 1978; 4: 1–7.
5　Nakata S, Petrie BL, Calomeni EP, Estes MK. Electron microscopy procedure influences detection of rotaviruses. Microscopy. *J Clin Microbiol* 1987; 25: 1902–6.
6　Suzuki H, Chen GM, Hung T. Effects of two negative staining methods on the Chinese atypical rotavirus. *Arch Virol* 1987; 94: 305–8.
7　Wood DJ, Bailey AS. Detection of adenovirus types 40 and 41 in stool specimens by immune electron microscopy. *J Med Virol* 1987; 21: 191–9.
8　Roberts IM. Immunoelectron microscopy of extracts of virus infection plants. In: Harris, JR, Horne RW, eds: Electron microscopy of proteins, 5. Viral structure. London: Academic Press, 1986.
9　Caul EO, Appleton H. The electron microscopical and physical characteristics of small round human faecal viruses; an interim scheme for classification. *J Med Virol* 1982; 9: 257–65.

Further reading

Microscopy

Chescoe D, Goodhew PJ. The operation of the transmission electron microscope. Royal Microscopical Society microscopy handbooks 02. Oxford: Oxford University Press, 1984.

Meek GA. Practical electron microscopy for biologists, 2nd edn. Chichester: John Wiley, 1977.

Weakley BS. A beginner's handbook in biological transmission electron microscopy, 2nd edn. Edinburgh: Churchill Livingstone, 1981.

Virology

Doane, FW, Anderson N. Electron microscopy in diagnostic virology. A practical guide and atlas. Cambridge: Cambridge University Press, 1987.

Madeley CR, Field AM. Virus morphology, 2nd edn. Edinburgh: Churchill Livingstone, 1988.

Photography

Coote JH. Monochrome darkroom practice – a manual of black and white processing and printing. London: Focal Press, 1982.

Electron microscopy and photography. Eastman Kodak Company, 1973.

11 Mycology

DWR Mackenzie

Differential media are relatively uncommon in mycology; moreover, the range of media for isolation and identification in most microbiological laboratories is generally limited. For these reasons, quality control is normally restricted to the following procedures, for which suitable test species are indicated. Unless individual reference strains are specified, recent clinical isolates of the test species are considered appropriate.

Isolation media

1 Glucose peptone agar (Sabouraud's)
2 Malt extract agar
3 Dermatophyte test medium (DTM)

If obtained commercially, each batch should be tested for ability to support growth of *Trichophyton mentagrophytes* and *Candida albicans*. On DTM medium, colour change should be observed only with growth of *T mentagrophytes*.

Commercially-produced or 'home-made' isolation media usually contain antibiotics to suppress growth of unwanted bacteria and moulds. It should be noted that some pathogenic fungi and actinomycetes may also be inhibited; choice of media should therefore be influenced by the possible identity of the suspected pathogen. Details of isolation media are provided in most modern mycological handbooks.

Mycology

The antibiotics in general use are:

1 Chloramphenicol, for suppression of bacteria (0.05mg/ml)
2 Cycloheximide, for inhibiting contaminating filamentous fungi (0.05mg/ml).

For media containing antibiotics, recommended test isolates are shown in Table 11.1.

Identification media/procedures

Yeasts are generally identified by means of commercial kits, supplemented by additional morphological tests. Table 11.2 lists recommended procedures and quality control species for tests commonly used for the identification of pathogenic yeasts and moulds.

Serological tests

Serodiagnostic kits obtained from commercial sources always contain internal controls. Always observe manufacturers' instructions regarding storage conditions and use. Serodiagnostic reagents prepared by individual laboratories should be assessed for potency and efficacy against a designated reference antigen or antiserum. These can be specially prepared for this purpose, or obtained from a commercial supplier (see Appendix 11.1) or a reference laboratory.

Preservation methods for quality control strains of fungi

Yeasts are most easily preserved by lightly inoculating 2ml quantities of sterile distilled water in small screw-capped glass bottles, such as bijoux. These can be stored at room temperature. Check viability and preservation of distinguishing features annually. This simple system

Table 11.1 Quality control for media containing antibiotics

Antibacterial antibiotics (chloramphenicol)	Anti-mould antibiotic (cycloheximide)	Control species for	
		growth	no growth
+	+	*Candida albicans*	*Cryptococcus albidus*
+	−	*Cryptococcus albidus*	*E coli*
−	−	*C albicans*	not applicable

Quality control in the microbiology laboratory

Table 11.2 Quality control for identification procedures

Test	Medium	Recommended control species
Chlamydospore production	Cornmeal, rice infusion, etc	*Candida albicans* NCPF 3153
Germ tube	Serum and others	*Candida albicans* (positive)
		C parapsilosis (negative)
Colony pigmentation	Nigerseed	*Cryptococcus neoformans* (positive)
		Candida albicans (negative)
Urease production	Urease medium	*Cryptococcus neoformans* (positive)
		Candida albicans (negative)
Hair penetration	Autoclaved hair	*Trichophyton mentagrophytes* (positive)
		Trichophyton rubrum (negative)
Vitamin requirements	Defined media	
	nicotinic acid	*Trichophyton equinum*
	thiamine	*Trichophyton violaceum*
	histidine	*Trichophyton megninii* (or *T rubrum*)
	inositol	*Trichophyton verrucosum*
Colour and sporulation	Lactrimel	*Trichophyton rubrum* (colour)
		Microsporum canis (sporulation)

can also be used for preserving mould fungi such as *Trichophyton mentagrophytes* and other filamentous fungi, provided they are sporulating.

Other methods for preservation include storage of slope cultures at $-18°C$, overlaying with sterile paraffin oil, and freeze-drying. Recovery tests should be made at intervals of six months (for isolates stored at $-18°C$) or one year (for oil-layered and freeze-dried cultures).

Appendix 11.1   Addresses

Suppliers of serological reagents for mycotic diseases are:

Mercia Diagnostics
Mercia House
Broadford Park
Shalford
Guildford
Surrey GU4 8EW
UK

Mycology

Northumbria Biologicals Ltd
South Nelson Industrial Estate
Cramlington NE23 9HL
UK

Alpha Laboratories
40 Parham Drive
Eastleigh
Hampshire SO5 4NU
UK

Suppliers of cycloheximide are:

ER Squibb & Son Ltd
Squibb House
141–149 Staines Road
Hounslow
Middlesex TW3 3JB
UK

12 Parasitology

AH Moody, L Odwell and PL Chiodini

The present practice of parasitology as a routine discipline relies heavily upon observer experience to detect and recognize the characteristics of a particular parasite. Quality control within the discipline is necessary to ensure reproducibility and sensitivity of detection. It requires strict control of techniques and apparatus and observer training.[1,2]

The microscope

The choice of a microscope suitable for diagnostic work is critical. The microscope should be binocular and ideally fitted with wide-field eyepieces of 10× power. A mechanical stage is essential and the instrument should have a moveable sub-stage condenser and an iris diaphragm for critical control of light entry. A flip-top secondary condenser lens should be present for high power work to provide the correct intensity and refraction of light. The most suitable objectives are the 10×, 40× and 100× and should be of the flat-field planachromatic type.

Calibration

Measurement is perhaps the most important criterion used for differential diagnosis in parasitology as all protozoan cysts and many helminth ova and other objects require differentiation by measurement of size. This is achieved by using a micrometer situated in the microscope eyepiece. The graticule is calibrated for each objective of the microscope by use of a micrometer slide. Once calibrated, a record is made of the size in micrometers corresponding to an individual division of the eyepiece graticule for each objective used. Recalibration is unnecessary. It is useful to keep the micrometer in a separate eyepiece, which can be rapidly inserted into the microscope when needed.

Quality control for detection of faecal parasites

Collection and delivery

If dysentery due to a parasitic organism is suspected, delivery of the faecal sample to the laboratory and its examination within 15–20 minutes is essential. Many protozoan trophozoites will survive longer than this, but all are sensitive to temperature and lack of moisture. In temperate climates preparations should be made in warm (37°C) 0.9% sodium chloride solution. Preservation of faeces in 10% formalin or another suitable preservative may be required for postal samples.

Concentration of faeces

For the examination of faeces which are not diarrhoeic and which may contain ova and cysts at low density, reliance is placed on concentration methods, eg the formol-ether technique, to detect parasites.[3]

The formol ether method involves several steps which should be standardized.

Formalin solution   The concentrated formalin solution (37% formaldehyde) should be diluted with distilled water (1 part formalin plus 9 parts distilled water).

The sieve   The sieve should be made either of brass wire with a pore size of 400–450μm, or of nylon mesh with a pore size of 600 $\times$ 450μm. Variation above or below this range will either cause over-retention of parasites on the filter or allow too much debris through.

Cotton gauze does not provide a satisfactory method of filtration due to the problems of irregular hole size and adsorption of material to the gauze.

Quality control in the microbiology laboratory

The centrifuge The centrifuge should be calibrated to provide a force of 1000G (RCF = $11.18 \times 10^{-7} \times RN^2$ when R = radius in mm from centrifuge spindle to tip of tube and N = speed of spindle in RPM). For hand centrifuges or low voltage centrifuges, a time calibration using known samples is required.

The use of stains

With wet preparation Stains provide valuable help in the identification of protozoan cysts in a wet preparation.

Each stain must be regularly checked against cysts of known identity to ensure correct performance.

Lugols iodine/acetic acid can be tested using two cysts, one of which should be *Iodamoeba sp*, the glycogen vacuole of which stains dark brown with iodine. The second should be an *Entamoeba* sp to confirm nuclear characteristics. Specific DNA stains (Burrow's, Sargeunt's) require fresh young cysts of *Entamoeba histolytica* to control the staining of chromatoid bodies.

With faecal smears Batches of faecal smears to be stained by auramine, modified Ziehl Nielsen methods or monoclonals for Coccidian öocysts should be controlled with known positive smears. Control faecal smears can be stored at 4°C after fixation with methyl alcohol for 15 minutes.

Faecal smears to be stained by the rapid Field's stain or Giemsa method cannot be controlled unless faeces containing fresh protozoan trophozoites is available. However, a blood film fixed in methyl alcohol for three minutes can be stained by these methods to check for correct staining of erythrocytes and leucocytes.

Preservation of faecal parasites for quality control

Methods used for the preservation of standard faecal material for use in the evaluation of a new or modified procedure, for an exercise of technical instruction or for a test of competence should preserve ova and cysts so that their diagnostic morphology is retained as nearly as possible. Some preservatives alter the morphology of parasites by collapsing cysts or allowing ova to continue development, while others may alter the behaviour of cysts and ova in concentration methods.

We have found Bayer's solution to be the most suitable for long-term preservation. When stored in a 1/10 dilution of the stock solution, ova and cysts maintain their morphology well over long periods and do not alter their concentration characteristics by standard formol-ether techniques. The formula for Bayer's stock solution is: $CuCl_2$ 7g, 20% V/V formaldehyde 1l, glacial acetic acid 70ml.

Reference material

For the provision of a reference collection and teaching material there is no real substitute for the preparation of permanent slides, especially of unusual or uncommon parasites. For long-term preservation, the most useful method is glycerine jelly mounts from formalized deposits which have been processed in glycerine/alcohol. Such slides will keep for several years.

Blood parasites

Microfilaria

Variables which can affect the detection and correct identification of microfilariae isolated from blood by the filtration technique include the following.

The time that the blood is collected Unless a particular species of known periodicity is sought, blood should be collected at midday and midnight.

The anticoagulant used Acid-citrate-dextrose or sodium citrate is recommended.

The choice of polycarbonate membrane Filtration of microfilaria requires a round pore of 3–5µm diameter. A number of membranes with irregular pores are also available but are not so satisfactory. Nuclepore 5µm 25mm diameter (Sterilin Ltd) have been found satisfactory. Skin microfilariae do not exhibit periodic patency. Skin snips should be incubated for up to four hours in 0.9% sodium chloride solution to allow time for the microfilariae to escape from the tissue.

Malaria parasites

Optimal diagnosis of malaria requires a collection of blood at a time when the parasitaemia is highest, although sampling and, if appropriate, treatment must not be delayed for this reason. Parasites must be well stained for correct speciation and accurate parasite counts.

Collection   The chances of finding malaria parasites are increased if blood smears are collected as near as possible to the time of peak density as indicated by rigor and fever. However, parasites may be present even in the absence of fever, so if a clinical suspicion of malaria exists, a blood film is mandatory, whatever the patient's temperature.

Giemsa staining of malaria parasites in blood smears at a pH of 7.2 provides the best conditions for demonstration of parasite structure and the presence or absence of red cell inclusions. Each new batch of stain should be tested using freshly-prepared blood films containing *Plasmodium vivax*. In the absence of fresh material, films stored with silica gel in a sealed air tight container at 4°C may be used. Failure to demonstrate Schuffner's dots might lead to misdiagnosis, and if a stain fails to achieve this on a known *P vivax*, it must be discarded. The quality of Field's stain is controlled by using freshly-prepared thick blood films, although unfixed thick blood films well wrapped in tissue and plastic film can be stored at −20°C for long periods and will provide an alternative source of control material.

Morphological variation of malaria parasites and exflagellation of gametocytes occurs on prolonged exposure to EDTA anticoagulants and must be avoided. Such changes can occur as early as two to three hours after collection, so prompt delivery to the laboratory is essential.

Other parasites

Giemsa staining of *Leishmania* sp requires stain buffered at pH 6.8.

In addition to the specific measures detailed above, high-quality diagnostic parasitology requires patience and application on the part of the observer.

References

1   Moody AH. The development of internal quality control and external quality assessment in parasitology. Developing countries proceedings. Stockholm: IAMLT Congress, 1986.

2   Fleck SL, Moody AH. Diagnostic techniques in medical parasitology. Bristol: Wright, 1988.

3   Allen AVH, Ridley DS. Further observations on the formol ether concentration technique for faecal parasites. *J Clin Path* 1970; **23**: 545–6.

13 Water microbiology

AE Wright

Quality control of water microbiology should not be confined to the technical examination to which a sample is subjected in the laboratory. It is unfortunate but true that the methods of examination available are subject to operator error and the results to wide statistical variation. It is essential, therefore, that the laboratory worker should be satisfied that the specimen is collected in a satisfactory manner, in a proper container, and transported to the laboratory within six hours under ideal conditions. It is also important that the individual who receives the report should be aware of the statistical problems inherent in the method.

It follows from this that the QC of water microbiology should be considered under these headings:

1 Specimen collection
2 Technical methods
3 Interpretation of results.

Specimen collection

In the UK the collection of specimens of water for bacteriological examination is usually done by environmental health officers. These officers, if in touch with their local laboratory, should be aware of the underlying philosophy of the method of examination, although they are often not aware of the limits of accuracy of the test. Specimens are usually collected at regular intervals as stipulated by the European Community, but others may be collected as a follow-up of a failure or as a consequence of a consumer's complaint. Clearly, these different types of specimen may require different treatment and there is therefore a need for adequate information to accompany each specimen. The information should state the type of water, whether raw or treated, the reason for sampling and the source, the date and, if treated, an estimate of chlorine present. Other useful information might include comments on recent weather, signs of visible pollution and, of course, the name of the person to whom the report should be sent.

Although in the past, collection bottles were always of good quality glass with ground glass stoppers, it is now commonplace to use plastic containers with screw-caps. The bottles should contain sodium thiosulphate at a concentration of 18mg/l, enough to neutralize up to 5mg/l of free and combined residual chlorine. It is usual to add 0.1ml of a 1.8% (w/v) solution of sodium thiosulphate for each 100ml of specimen.[1] This will not affect *E coli* or coliforms in untreated water, as long as the time taken to reach the laboratory[2] does not exceed six hours.

The solution of sodium thiosulphate and indeed of other chemicals and media should be made with sterile distilled water. The purity of the water should be checked from time to time using a conductivity meter. This should not exceed 1.5ppm of NaCl and should preferably be below 1.0.[3]

Technical methods

The basic philosophy of water bacteriology is based on the fact that it is impracticable to search for the presence of pathogenic organisms in large quantities of water. This is because pathogens such as *Salmonella typhi* tend to die out rapidly in water, and in any case are likely to be vastly outnumbered by other bacteria commonly present in human and animal excrement. Thus there is no need to seek for pathogens when these other 'indicator' organisms are more easily detected. These indicator organisms include coliforms and *E coli*, as defined by water bacteriologists in a pragmatic definition not entirely in favour with taxonomists. Other organisms indicative of faecal pollution include faecal streptococci and the spore-bearing organism, *Clostridium perfringens*. The former are outnumbered by the coliform group in sewage while the latter may persist for long periods and be indicative of faecal pollution in the more distant past. Other organ-

Water microbiology

isms present in water may cause complaints due to discolouration or to a reduction in the shelf life of food or milk but can be detected by simple plate counts. Each of the organisms mentioned can be detected relatively quickly by methods developed over the years by water bacteriologists. All the methods used should be subject to quality control. The present indicator organisms will not cover all aspects of microbiological safety; for example, cryptosporidium and giardia may need to be borne in mind in the development of water microbiology QC. Differential resistance among microorganisms may be used to advantage, eg cryptosporidia are resistant to chlorine compared with other organisms.

The principal methods used to detect the presence of indicator organisms are the multiple tube method and the membrane filtration technique.

Multiple tube method

In this method, measured quantities of water are added to tubes containing predetermined amounts of selective media. These tubes are incubated and on the assumption that one organism is sufficient to seed a tube and to grow, so a presumptive count of organisms can be obtained by consulting probability tables. For example: if three sets of five tubes are inoculated with 10ml, 1ml and 0.1ml volumes of water and one tube in each set shows acid or gas production, then the mean probable number of coliform organisms present will be six. This finding must, of course, be confirmed by further tests. As the procedure is capable of detecting as few as one coliform or *E coli* in 100ml of water, and as the presence of one *E coli* is sufficient to class a supply as unsatisfactory, QC is clearly important.

The general QC tests to which media should be subjected are dealt with elsewhere (see Chapter 2) and will not be repeated here. The regular, repeated use of control organisms is important to ensure that the production of acid and gas in the liquid medium used is consistent. In this connection the temperature of the water baths and incubators used may be crucial.

Detection of coliform organisms is usually done at 35°C or 37°C, sometimes with a preincubation for four hours at 30°C to allow resuscitation of damaged organisms. Confirmatory tests for *E coli* should be done in a water bath, controlled to give a reading of 44 ± 0.25°C *in the medium*. A rise of only 1°C is sufficient to inhibit gas

Quality control in the microbiology laboratory

production of certain *E coli* and a fall to 42°C will allow other coliforms to give false positive results. The best way of ensuring these temperatures is to use a thermometer conforming to British and/or international standards (see Appendix 13.1). Readings should be taken in a tube of media in the water bath and observations recorded in a workbook daily.

Membrane filtration technique

When the membrane filtration technique is used, the 47mm-diameter filters should have a pore size of 0.45μm and should be marked with a grid to make counting easy. Control organisms (*E coli*, positive and *Klebsiella aerogenes*, negative) should be used frequently to ensure the suitability of the membranes and that the material used to mark the grid does not inhibit the growth of these controls.

If membranes are sterilized in the laboratory, they can be autoclaved at 115°C for 10 minutes or boiled in distilled water for the same time. Excess heat and careless handling may affect the efficiency of membranes and they should always be examined for damage prior to use.

As with the tube method the temperature of water bath and incubator is important. If pre-incubation is done in an incubator it should be anhydric, as a water-jacketed incubator takes too long to stabilize.

When preincubation at lower temperatures is followed by an automatic switch over to a higher temperature, the former should be at least 5°C above ambient.

When dual-temperature water baths are used, it is most important that the watertight container in which the petri dishes are housed is loaded in such a way that the dishes are below the surface of the water. Failure to observe this point will lead to false results.

The rate of flow through a membrane filter may be of importance and some authorities therefore recommend slow filtration using a vacuum of 500mm of Hg. It is also important to ensure an even distribution of colonies over the surface of the membrane to prevent crowding and to facilitate counting. This is best done when filtering raw and probably polluted waters by running sterile distilled water through the membrane first or by diluting the water prior to filtration. In both the multiple tube method and the membrane filtration technique, two other aspects of QC are important. The first concerns the mixing of the original specimen.

Water microbiology

Microbiologists are well aware that organisms tend to clump and some to adhere to glass surfaces. When one is seeking to demonstrate one organism in 100ml of water, it is clearly important that mixing should be adequate. An older generation of microbiologists recognized this and stipulated[4] that 'the sample bottle should be inverted 25 times by a rapid rotary movement of the wrist', and after pouring off a little fluid the bottle should be shaken again. This advice, in a recent edition of the same report,[1] now reads, 'Invert the sample bottle rapidly several times'. Although the importance of mixing is referred to elsewhere in the report, it is possible that the result of the investigations may hinge on the operator's interpretation of the word 'several'.

The last point concerns care in the measurement of the quantities of the sample dispensed. Pipettes should be of good quality glassware and thoroughly washed prior to sterilization.

Interpretation of results

It is not often realized that the multiple tube method for the examination of water has a large sampling error. Indeed, the upper limit could be as high as three times the reading given in the tables, while the lower limit may be between one third and a quarter. In 5% of the estimations, the true figure may even lie beyond these limits.[5] A brief description of the statistical considerations and a list of further references can be found in Report 71.[1]

It is clear, therefore, that care should be taken when comparing results between two specimens. For this reason, samples of water supplies should be taken and examined frequently to establish a norm. The single isolated sample may be inaccurate and in any case can only apply to the water supply at that particular time, a time which may well have been influenced by local weather conditions.

The membrane filtration technique is also subject to statistical variation. From a colony count of C the following formula gives the 95% confidence limits.

$$C + 2 \pm 2\sqrt{C + 1}$$

Counts of less than 20 are even more inaccurate – thus a count of 10 has limits of from three to 18. In addition, membrane filtration is known[6] to give lower counts and more false negative results with *E coli* counts than the multiple tube method. False-positive reactions

Quality control in the microbiology laboratory

due to the growth of aerobic or anaerobic spore-bearing organisms occasionally encountered with liquid media do not occur with membranes.

Quality control of the various processes outlined above should be augmented by regular participation in external assessment exercises. Results from such QA specimens in the UK have shown that laboratories are capable of producing consistent results within the statistical limits described above. Such results are probably due to adequate internal QC procedures.

References

1 Department of the Environment, Department of Health and Social Security and the Public Health Laboratory Service. Report 71: The bacteriological examination of drinking water supplies. London: HMSO, 1982.

2 PHLS. The effect of storage on the coliform and *Bacterium coli* counts of water samples: storage for six hours at room and refrigerator temperatures. PHLS water sub-committee. *J Hyg* 1953; **51**: 559–71.

3 Porton IR, Brown R, Wilkinson JF. pH measurements and buffers, oxidation-reduction potentials, suspension fluids and preparation of glassware. In: Collee JG, Duguid JP, Fraser AG, Marmion BP, eds: Practical medical microbiology (Mackie and McCartney). Edinburgh: Churchill Livingstone, 13th ed, 1989.

4 Ministry of Health. Ministry of Housing and Local Government. Report 71: The bacteriological examination of water supplies. London: HMSO, 1956.

5 Swaroop S. The range of variation of the most probable number of organisms estimated by the dilution method. *Ind J Med Res* 1951; **39**: 107–34.

6 Tillett HE, Wright AE, Eaton S. Water quality control trials: statistical tables for direct comparison between membrane filtration bacterial counts and the multiple tube method with a description of the bacteriological method. *Epidemiol Infect* 1988; **101**: 361–6.

Appendix 13.1

Standards for thermometers

1 ISO 1770 Solid stem general purpose thermometer 0–60°C Scale divisions 0–1°C 1st ed. 1981/06/15 International Organization for Standardization, Geneva, Switzerland
   or
2 BS 1704/1951 updated to 1985. This standard is now in accord with the international standard.

14 Food microbiology – a PHLS perspective

MH Greenwood and WL Hooper

Food microbiology exemplifies the interface between clinical and industrial microbiology. Laboratory aspects of quality control in the investigation of food in suspected food poisoning do not differ from those applicable to clinical specimens. A search for the causative organism requires selectivity of cultural techniques and precision in identification, although sensitivity of detection is normally of less importance, since the agent responsible for causing the symptoms will be present in large numbers in the majority of cases. There are occasions when sensitivity is important, however, and the methodology must be suitable to the circumstances and nature of the food.

Routine examination of food and environmental specimens from premises where food is prepared, sold or handled demands a more specialized approach. Here the emphasis must be on established criteria and deviation from the norm. Food is now available in a vast array of presentations, ranging from raw foods to materials which have been subjected to various types of preservation, heat treatment and processing to sterility. Laboratory examination therefore requires emphasis on precision of counting organisms for comparative purposes, linked with detection of pathogens, potential pathogens or indicator organisms present in small numbers in defined quantities of material. Attention to detail in reporting is essential. A report stating 'no pathogens isolated' is meaningless, unless there is a clear understanding of what has been looked for, the quantity actually subjected to test, and the sensitivity of the test methods. Extreme care must be exercised in interpreting laboratory findings unless precise details are known of the history of the previous handling, processing and storage of the food sample. Laboratory assessment can often be made much easier if the food is known to come from a manufacturing plant where hazard analysis of critical control point (HACCP) principles and practices are employed. Knowledge of the weaknesses in the food chain prior to receipt in the laboratory goes a long way towards ensuring the value of test results. Different criteria must be applied to foods at the point of production, point of sale and point of consumption.

Microbiological examination of foods and food products is usually undertaken in order to monitor the hygiene of processing, catering or retailing practices, to assess the quality and potential shelf life of the product, or to detect the presence of pathogenic or toxigenic organisms which may give rise to illness. Hygiene, quality and shelf life are usually assessed on the basis of the total numbers of viable bacteria present and also the presence and number of certain indicator bacteria such as coliforms. Reliable estimations of bacterial counts can only be attained by strict control of the test methodology and counting procedures used. In contrast, enteric pathogens such as salmonella or campylobacter should not be present in food which is ready to eat, and examination for the presence of pathogens is carried out on a basis of presence or absence in 25g or more of food. Confidence in a negative result requires the knowledge that the method used for isolation of the pathogen is capable of detecting that organism, which may have been stressed or sub-lethally injured by the food processing. Use of reference samples analysed by different laboratories in comparative trials has shown that variation in results obtained from the same sample may be due to a lack of homogeneity of the sample, to differing storage conditions prior to examination and to differing sampling procedures and techniques of analysis.[1] Implementation of the aspects of quality control discussed in this chapter will help to ensure repeatability (on testing within the same laboratory) and reproducibility (on testing among different laboratories) when examining food samples.

Sampling

Ideally the number of samples which should be tested depends upon the degree of hazard to the consumer and conditions of use of the

Food microbiology – a PHLS perspective

food item. The principles and applications of sampling for microbiological analysis and the statistical considerations of sampling have been discussed by the International Commission on Microbiological Specifications for Foods.[2] The food sample should be collected either in the original container or aseptically transferred to a sterile container.

Transport and storage of food before examination

Perishable food should be placed immediately in an insulated cold box and transported rapidly to the testing laboratory. On arrival the sample should be transferred to a 4°C refrigerator (or freezer for frozen food) and not removed until testing is about to commence. Frozen items should be defrosted at 4°C. After examination the food should be stored again at 4°C. Non-perishable samples should be stored in a cool, dry area.

Preparation of food sample for colony counts

Aseptic techniques should be used throughout the handling of the food item. These may include swabbing the outer wrapping of the food with 70% alcohol to prevent contamination of the food as the test portion is removed. A portion of at least 10g is weighed, using a balance with a sensitivity of 0.1g with a 200g load. A taring facility on the balance is desirable. If the sample is liquid, it should be mixed by inverting the container 25 times, and a 10ml test portion removed as soon as possible after mixing. Sufficient sterile diluent is then added to make a 1:10 dilution. The diluent should be non-toxic; a suitable diluent for most food materials is peptone/saline solution (0.1% peptone, 0.85% saline).[3]

Microorganisms in food are rarely distributed in a homogeneous way, and may be trapped within the food. In order to release them into the liquid diluent and to help break up clumps of bacteria, the test portion should then be homogenized using a stomacher for 30–60 seconds, or in a sterile blender for one to two minutes. Excessive blending should be avoided to prevent injury to the organisms due to heat generation or the mechanical action of the blender. Further dilutions may then be prepared from this homogenate. The use of automatic pipettors is recommended to ensure accuracy. Thorough mixing is required as each dilution is prepared; this may be achieved by aspirating 10 times with a pipette, shaking 25 times through a

25cm arc, or by use of a vortex mixer. The homogenate and these dilutions should then be used for the relevant counting techniques with the minimum delay, not exceeding 20 minutes.[4,5]

Colony counts

There are several methods suitable for performing colony counts. (A guide to these methods is in preparation.[6]) All counting methods should be defined in terms of the medium used and the length and temperature of incubation. A total viable count (aerobic plate count, standard plate count) is normally performed using plate count (standard methods) agar;[7] the addition of skimmed milk powder is recommended for dairy products.[3] These media are translucent, and can be used for all counting methods, thus improving comparability of methods. counting methods, thus improving comparability of methods. Incubation may be carried out at various temperatures, depending on the food product and the information sought. Plates from dairy products are incubated at 30°C, and this temperature is also widely used for enumeration of mesophilic bacteria in other foods. Incubation at 35–37°C is also used for counting mesophilic bacteria. Other temperatures such as 6°C or 55°C are chosen for enumeration of psychrophiles and thermophiles. The length of incubation should also be defined, for example 48 ± 2 hours or 72 ± 2 hours. Statements specifying 'overnight' incubation or 'incubation for two days' are not specific enough when trying to obtain precise, reproducible counts. If plates are removed from the incubator before the end of the incubation period, they should be reincubated as soon as possible, and not left at ambient temperature for any significant length of time. The temperature inside the incubator should be checked daily, and maintained to within 1°C of the specified incubation temperature. Opening and closing of incubator doors should be kept to a minimum. Plates should be inverted (agar surface facing downwards) and placed in the incubator so that the agar equilibrates to incubation temperature within two hours. Overcrowding should be avoided and dishes should not be stacked more than six deep.[8]

The most commonly used methods for performing total viable counts are the pour plate, surface drop, surface spread and spiral plate.[7,9] Each method has its advantages and disadvantages, and none of the methods can be relied upon to enumerate all types of

Food microbiology – a PHLS perspective

organisms present. A method has been described for establishing acceptability of alternative counting methods when compared with a standard method using the same samples of food.[10] Samples are plated in duplicate using both methods until 25 samples yield plate counts of 10–300 colonies by both methods. For each method:

1 Duplicate counts are converted to log_{10} values.
2 The difference between the log_{10} counts for duplicate plates is squared.
3 The squared values are added for all samples.
4 The log_{10} counts are added together and divided by 50 to obtain the mean log_{10} count.

The mean log_{10} count for the 50 observations using the alternative method should not differ from the same value for the standard method by more than 0.036. For satisfactory duplicate data, the sum of the squared differences (log variances) between duplicate samples should not be greater than 0.005.

The pour plate (or standard plate count) method is frequently stipulated in reference methods; if this method is used the most important factor for control is the temperature of the molten agar, which should not exceed 45–46°C.[5,7,11] The agar should be used within three hours of melting. Control plates should also be prepared to check sterility, one containing agar only and another containing agar and 1ml diluent. Solidified plates are used for surface counting techniques, and these should be level and free of air bubbles. They may require drying before use so that the inoculum is absorbed within 15 minutes; drying at 50°C for 1.5–2 hours has been recommended.[7] The plates should be allowed to reach room temperature before inoculation. If a spiral plater is used, the instructions given by the manufacturer should be followed to ensure even plating; in addition the starting position and lift-off point of the stylus should be checked daily and adjusted if necessary.

Results obtained by use of surface counting methods may be less precise than the pour plate method as smaller volumes of the 10^{-1} homogenate and subsequent dilutions are generally used; this is particularly true of the surface drop and spiral plate methods when colony counts are less than $10^3/g$. Surface spreading of a 0.5ml volume will improve precision when counts are low. For improved accuracy, duplicate plates for each dilution should be prepared.

Quality control in the microbiology laboratory

Counting of colonies and expression of results

All pour plates or surface spread plates containing 10–300 colonies, or surface drops with up to 20 colonies per drop, should be used for counting.[7,11] Spiral plates should be counted and results computed according to the manufacturer's instructions. Colony counts should be performed within four hours of the end of incubation, or if this is not possible, after they have been stored overnight at 0–5°C. If overgrowth occurs, due for example to spreading colonies of *Bacillus* species, count the colonies in the half of the plate that is clear and multiply by two. Reject any plate in which more than half of the plate is overgrown.

Analysts should be able to duplicate their own counts on the same plate within an 8% variation, and the counts of other analysts within 10%, on 90% of samples.[12] If automated colony counters are used, they should yield counts that are within 10% of those obtained manually on 90% of samples.[12]

The number of microorganisms, N, per g or ml of sample can be calculated from the following formula.[13]

$$N = \frac{C}{V(n_1 + 0.1n_2)d}$$

where C is the sum of colonies on all plates counted
n_1 is the number of plates in the first dilution counted
n_2 is the number of plates in the second dilution counted
d is the dilution from which the first counts were obtained
V is the volume applied to each plate.

Example

Volume applied 1ml
Dilution 10^{-2} 282 and 293 colonies
Dilution 10^{-3} 27 and 32 colonies

$$N = \frac{282 + 293 + 27 + 32}{1(2 + 0.1 \times 2)10^{-2}}$$

$$= \frac{634}{0.022}$$

$$= 28,818 \quad \text{expressed as } 2.9 \times 10^4$$

In reporting the result, round off the number to two significant figures and express as a power of 10. In the example above, the result would be rounded off to 29,000 and expressed as 2.9×10^4 colony forming units (cfu) per g or ml. When the digit to be rounded off is 5 with no further significant figures, round off so that the figure immediately to the left is even, eg 28,500 is rounded off to 28,000.

If less than 10 colonies are produced in pour plates or surface spread plates of the lowest dilution used, report the number of microorganisms as less than $10 \times d$ cfu per g or ml, where d is the reciprocal of the lowest dilution factor. If all plates fail to produce colonies, report the number of microorganisms as less than $1 \times d$ cfu per g or ml.

Detection and enumeration of specific organisms

Specific organisms such as *Staphylococcus aureus*, *Clostridium perfringens* and *Escherichia coli* are often present in low numbers in foods, and so the 10^{-1} homogenate is usually used for their enumeration. Specialized selective agar media have been developed for the isolation and presumptive identification of each individual organism or group of organisms, and once again, duration and temperature of incubation are critical factors. Confirmation of presumptive positive colonies should be performed on five colonies of each type, or all colonies if less than five are present.[14] If demonstration of the presence or absence of these organisms is required, a known weight (eg 1g, 10g) of the food is added to a suitable selective liquid medium. Enumeration can also be carried out in liquid media, using a most probable number (MPN) method.[7,14,15] If the detection of a presumptive positive tube depends upon gas production, then care must be taken to ensure the absence of air in the inverted Durham fermentation tube before incubation. Confirmation of positive tubes is usually necessary to ensure that the reaction has been produced by the organism sought by breakdown of the medium and not the food. A guide to methods for the detection and enumeration of specific organisms is in preparation.[6]

The selective media most commonly used in the analysis of foods are listed elsewhere.[16] As a general rule, recovery on the selective agar medium should be within 0.5 log_{10} of the counts obtained on a non-selective agar, and the selective medium should aim to reduce other competing organisms by 3–5 log_{10}. The MPN count obtained from a selective broth should not be less than 70% of that obtained in a non-selective broth.[16]

Resuscitation of injured organisms

Most foods have been processed in some way, and this may cause damage to the surviving organisms, for example by the effects of heating, freezing, drying, low pH or high salt content. Techniques used during analysis, such as heat stress caused by molten agar used in the pour plate technique, may inflict further damage. The choice of method for enumeration or isolation of specific organisms should take account of the type of food to be examined and allow for recovery of stressed organisms. The recovery of these sub-lethally injured organisms may necessitate a resuscitation stage prior to their exposure to selective agents such as bile salts, antibiotics and high temperatures of incubation. The optimum temperature (25°C–37°C) and the time required (one to four hours) for repair vary according to the nature of the stress.[17] This recovery stage involves the use of non-selective media, with addition of or transfer to appropriate selective media after the repair period. Some selective media contain components which aid repair, recovery and growth, such as sodium pyruvate which degrades hydrogen peroxide, or magnesium chloride, to replace that lost due to ribosome and cell membrane injury.[18] These factors should also be borne in mind when choosing the selective medium for the detection of a specific organism. The efficacy of a resuscitation procedure may be assessed by subjecting a suitable culture of bacteria to artificial stress such as freezing to $-20°C$ for 24 hours or heating to 50°C for 30 minutes, and comparing recovery after resuscitation with that of a non-stressed culture.

Detection of pathogens

The presence of pathogens such as salmonella and campylobacter in processed foods is considered significant, regardless of the number of these organisms. Due to the effects of processing, these organisms may be present in very low numbers. Direct enumeration is therefore rarely attempted, and the presence of pathogens is usually sought by enrichment of a significant quantity of food (eg 10–100g). Because of the likelihood of injury, and in order to increase the number of organisms present before exposing them to selective agents, a pre-enrichment stage in a non-selective medium is frequently incorporated. The length of incubation of this stage is critical, and must allow time for the resuscitation and multiplication of low numbers of the pathogen to a level which may be detected if only an aliquot is then used for selective enrichment. For example, best recovery of salmonella is obtained after incubation of the pre-enrichment medium for 18–24 hours. Liquid media used in the isolation process should be allowed to reach ambient temperature before use to avoid cold stress.

The performance of an entire isolation procedure such as that used for the recovery of salmonella can be checked by monitoring the recovery of low numbers of an added test organism. The organism may be inoculated into an appropriate foodstuff previously shown to be free of that organism, or it can be inoculated directly into the liquid medium along with cultures of organisms representing the normal flora which might compete in a sample of food.[16] The isolation method should be capable of detecting an inoculum of 20 or fewer organisms.[16] In practice the detection of 1–5 cells of salmonella in 25g of food can be achieved.

Evaluation of the suitability and effectiveness of a procedure may be carried out as follows. An overnight broth culture of the test organism is diluted 10^{-1} to 10^{-9} in peptone saline solution. The viable count is estimated by surface plating a known volume of each dilution on a non-selective agar such as tryptone soya agar. A 1ml volume from each of the four highest dilutions is inoculated into separate homogenates of 10g of food in 90ml of pre-enrichment or enrichment broth, and subjected to the full isolation process. The highest dilution used from which the test organism was recovered is determined, and the minimum number of organisms required to initiate growth is calculated from the viable counts obtained previously.

The use of automated instruments and kits

Microbiological analysis of foods is time consuming both in terms of technician time and total elapsed time before a result is obtained. In order to reduce these times, the food industry is turning to the use of automated instruments which measure changes in the electrical characteristics of a medium as microorganisms grow (eg impedance, conductance). These electrical changes occur when a threshold level of 10^6–10^7 organisms per ml is reached,[10] and a bacterial growth curve is produced when the electrical changes are plotted as a function of time. The time required to reach the threshold level (detection time) correlates with the original number of bacteria present. The growth medium may be made selective in order to detect specific organisms such as coliforms or salmonella. The most important aspect of control is the preparation of specific product calibration curves, which will vary for different food commodities, by performing a series of parallel samples to relate the detection times to bacterial counts obtained by routine plate count procedures. Detection time 'cut-off' levels can then be established which determine whether the bacterial load of a specific sample is above or below a given specification, or whether a specific organism such as salmonella is present. The accuracy and precision of the results depends on the culture medium used, which may need to be adapted for use with these instruments. Sterility of the medium should also be checked.

The use of kits for the detection of pathogens such as salmonella or listeria and for detection of toxins such as staphylococcal enterotoxin in food is also becoming more commonplace. The preparation of the food sample for use in kits may need to be varied according to the type of food, and non-specific results due to food components may be encountered. The components of the kit should be brought to room temperature before use, and the expiry date of the reagents checked. As always, positive and negative controls should be performed. When possible, a positive result should be checked by conventional methods. In addition, the user should be aware of the limitations of the test when interpreting the results.

References

1 van Leusden FM, van Schothorst M, Beckers HJ. The standard salmonella isolation method. In: Corry JEL, Roberts D, Skinner FA, eds. Isolation and identification methods for food poisoning organisms. London: Academic Press, 1982: 35–49.

2 International Commission on Microbiological Specifications for Foods (ICMSF). Microorganisms in foods 2. Sampling for microbiological analysis: principles and specific applications, 2nd edn. Toronto: University of Toronto Press, 1986.

3 British Standards Institution. BS 4285: Microbiological examination for dairy purposes, section 1.2. Diluents, media and apparatus and their preparation and sterilization, 1984.

4 British Standards Institution. BS 4285: Microbiological examination for dairy purposes, section 1.1. Sampling and preparation of sample, 1984.

Food microbiology – a PHLS perspective

5 Busta FF, Peterson EH, Adams DM, Johnson MG. Colony count methods. In: Speck ML, ed. Compendium of methods for the microbiological examination of foods, 2nd edn. Washington: American Public Health Association, 1984: 62–83.

6 Roberts D, Hooper WL, eds. Practical Food Microbiology. London: Public Health Laboratory Service. [In preparation].

7 International Commission on Microbiological Specifications for Foods (ICMSF). Microorganisms in Foods 1. Their significance and methods of enumeration, 2nd edn. Toronto: University of Toronto Press, 1988.

8 British Standards Institution. BS 4285: Microbiological examination for dairy purposes, section 1.3. Procedures for obtaining incubation conditions, 1984.

9 Gilchrist JE, Campbell JE, Donnelley CB, Peeler JT, Delaney JM. Spiral plate method for bacterial determination. *Appl Microbiol* 1973; **25**: 244–52.

10 Houghtby GA, Maturin LJ, Kelley WR. Alternative microbiological methods. In: Richardson GH, ed. Standard methods for the examination of dairy products, 15th edn. Washington: American Public Health Association, 1985: 151–71.

11 British Standards Institution. BS 4285: Microbiological examination for dairy purposes, section 3.14. Enumeration of mesophilic organisms, 1988.

12 Messer JW, Behney HM, Leudecke LO. Microbiological count methods. In: Richardson GH, ed. Standard methods for the examination of dairy products, 15th edn. Washington: American Public Health Association, 1985: 133–50.

13 British Standards Institution. BS 4285: Microbiological examination for dairy purposes, section 2.1. Enumeration of microorganisms by poured plate technique for colony count, 1984.

14 British Standards Institution. BS 4285: Microbiological examination for dairy purposes, section 3.7. Enumeration of coliform bacteria, 1987.

15 Oblinger JL, Koburger JA. The most probable number technique. In: Speck ML, ed. Compendium of methods for the microbiological examination of foods, 2nd edn. Washington: American Public Health Association, 1984: 99–111.

16 Baird RM, Corry JEL, Curtis GDW, eds. Pharmacopoeia of culture media for food microbiology. *Int J Food Microbiol* 1987; **5**: 187–299.

17 Ray B, Adams DM. Repair and detection of injured organisms. In Speck ML, ed. Compendium of methods for the microbiological examination of foods, 2nd edn. Washington: American Public Health Association, 1984: 112–23.

18 Hurst A. Revival of vegetative bacteria after sublethal heating. In: Andrew MHE, Russell AD, eds. Revival of injured microbes. London: Academic Press, 1984: 77–103.

15 Food microbiology – an industrial perspective

JA Bird

Microbiology laboratories handling food and environmental samples should be properly set up to receive and analyse these samples, and to evaluate and report the results. It is important that laboratory staff are trained in appropriate microbiological techniques in order to perform the duties assigned to them.[1] Work should be carried out according to procedures documented in a manual of methods, and the laboratory must have the correct facilities and equipment to enable personnel to carry out these procedures accurately and efficiently.

Food materials are very variable, both in composition and microbial content. Therefore, the analyst should have a knowledge of the specific storage conditions and preparation procedures appropriate to the particular food and its likely microbial flora when interpreting results. When performing microbiological examinations it is important to bear the following in mind:

1 Contaminants should not be introduced during the sampling or testing process so that only the microbes which are present in the sample are enumerated.

2 The microbes (which could be pathogenic) must not contaminate the environment. This could lead to the contamination of other laboratory samples or to laboratory personnel becoming ill.[1,2]

This requires particular emphasis on the use of hygienic and safe laboratory practices. For example, most sterilized canned products should be free of microbes capable of growth in the product during storage and should be opened in an environment free from microbial contaminants, such as a safety cabinet, to protect both the product and the laboratory worker. On the other hand a raw meat comminute may be expected to contain $c.10^6$ microbes/g and could be contaminated with a wide range of pathogens, so should be handled under conditions which minimize the contamination of the worker and environment, such as homogenization in a sealed container and avoidance of aerosols when pipetting etc.

Laboratory siting

The laboratory handling food samples should be purpose built for food analysis. It should be clearly identified as a food microbiology laboratory and be separate from all other working areas (including other laboratories). Personnel working in the food microbiology laboratory should be clearly identified and access to the laboratory should be restricted to these people only.

Laboratory practices

Good laboratory practices should be documented and enforced,[1,2] and must cover those included in the various health and safety regulations.[3,4,5]

Sampling

Correct sampling requires care to ensure that representative samples are taken for analysis; care should also be taken at all stages to avoid contamination and treatments which may affect microbial numbers. Personnel involved in sampling must be properly trained in appropriate sampling techniques and methods for labelling, storing and transportation of samples.[6]

Sampling equipment

Sampling equipment should be suitable for its intended purpose. Wherever possible, equipment used for removal of a sample of the food should be made of stainless steel or of another material of adequate strength that will not affect the microbiology of the food sample. For resterilizable equipment all surfaces should be smooth and free of crevices; all corners should be rounded to allow for easy cleaning and sterilization. All sampling equipment must be clean and sterile before use. Disposable presterilized equipment (including swabs) should be purchased from reputable sources.

Sample containers

All containers and closures should be dry, clean, leakproof and sterile. The shape and capacity of the containers should be appropriate to the particular requirements of the product being sampled.

Sample containers and closures should be of materials and construction which protect the sample and do not bring about a change in the sample which may affect the results of microbial analyses. For example, reusable containers may be adequately cleaned and sterilized but it is essential to avoid the carry over of cleaning chemicals which could either inhibit subsequent microbial growth or affect the selectivity of the analytical procedure. Additionally, the container closure should be capable of preventing sample contamination. Appropriate materials include some metals and plastics; but glass should be avoided when sampling foods in food premises. If transparent sample containers are used the filled containers should be stored in a dark place.

Disposable sterile plastic jars or bags (with suitable closures) are best used. It is good practice to place one bag inside another (double thickness) to resist puncture.

Sampling procedures

It is essential to sample aseptically, although with some products this may be difficult to achieve. Most sampling regimes are designed for randomly taken sample units. To overcome both of these problems specific sampling devices and schemes have been developed.[7] For instance, with large scale food production, in recognition of the difficulty of random sampling, frame or timed (continuous) sampling is practised. The size of the sample should be agreed by the parties involved. A knowledge of sampling procedures is essential for successful interpretation of the results.

Sampling should be carried out in accordance with the international standard appropriate to the product concerned. If there is no specific international standard, it is recommended that the parties concerned come to an agreement on this subject.[2]

The sample should be aseptically transferred to the sample container (as described above), sealed and clearly labelled.

Sampling plans

It is important when examining a food to apply a level of testing appropriate to the intended microbiological purpose, eg in a specific investigation or in a routine examination. A sampling plan will contain a statement or number of product units to be examined, the methods of analysis and the limits to be applied to each organism in the sampling plan. At its simplest, a sampling plan can be based on a single sample. However, such an analysis will be highly inaccurate because of the heterogeneous distribution of microorganisms in most foods. Thus, to improve the relability of decisions, multiple sample units are usually taken at random from a consignment of food. For routine acceptance testing of foods, most food microbiologists in the food industry apply attribute plans, such as those described by the ICMSF.[7] These plans provide details of the probability of acceptance of a batch of foods with defined levels of defects, and thus provide important information concerning the interpretation of the results of the analysis. For investigative sampling, where the purpose is to locate the likely cause of a problem, acceptance sampling is not appropriate and a variety of different sampling plans can be applied, depending on the particular investigation. This subject is complex but is well covered by Cochran.[8]

Sample labelling

Containers can usually be labelled by marking them with a felt-tip indelible marking pen. However, light-coloured waterproof cardboard tags with reinforced eyelet holes and wire or cord ties, gumbacked paper labels or adhesive-backed tape can be used to identify samples. Labels should be large enough to carry all relevant data.

The person responsible for taking the sample should also complete and sign a detailed sampling report, which should also be countersigned if representatives of other parties are involved. This report should contain the following information relevant to each sample:

1 Name, address and job title of the person collecting the samples.
2 Names and addresses of appropriate representatives of any parties involved.
3 Date, place and time of sampling.

Quality control in the microbiology laboratory

4 Reason(s) for sampling.

5 Nature of the food.

6 Names of manufacturer, importer, seller and buyer, as appropriate.

7 Size, number, and reference numbers of sample units.

8 Details of any markings on packages (eg use by date, manufacturer's code).

9 Temperature of the product at time of sampling (if appropriate).

10 Integrity of sample (packaging damaged or opened, product used).

11 Means of transporting samples to the laboratory and by whom; write any storage instructions on the outside of packages.

12 Name and address of laboratory analysing samples.

13 Analyses required (eg salmonella).

The report should also contain relevant information on factors, conditions or circumstances which are needed to enable correct selection of storage, preparation and analytical techniques.

Storage and transport of samples

Storage of samples before analysis should ideally be avoided, as this will inevitably affect the microbial population in some way. In practice, storage of samples is often necessary because of practical considerations. However, the storage period should be minimized.

Preservatives should not be added to any sample intended for microbiological examination.[6]

Swabs from surfaces etc may be stored in special transport or support media intended to protect stressed microbes until analysed. They should not be frozen; only qualitative analyses should be done on samples stored for longer than two hours.

When possible, the sample should be stored as it is. For example, dried powders are normally best stored as dried powders as changes in the flora are likely to be minimal. This is also true of many preserved, ambient stable foods. However, the microbial populations of wet preserved foods are often more likely to change with time. Similarly, some perishable foods may be preserved by storage at temperatures of 0–2°C for up to 24 hours, although in many such foods quite rapid changes in the microbial flora can occur at normal refrigeration temperatures, and therefore these types of foods are often stored and transported frozen. Generally, rapid freezing minimizes the changes in the microbial population, although some Gram-negative bacteria and vegetative cells of *Clostridium perfringens* may be destroyed by freezing and thawing in foods. Such freezing is often more efficiently carried out if the product is frozen as a relatively thin layer. Large samples may take some time to thaw, and thawing in a temperature controlled refrigerator at 0–5°C may be used to prevent excessive growth before analysis. Thawing times should always be as short as possible but should not be done under conditions which encourage growth in the thawed food. Frozen products can be stored in this state until required for analysis.[2]

It is essential to record all sample storage conditions if the results are going to be interpreted correctly. Such records should include the history of the sample, indicating any abuse. For example, a sample of a canned product taken two days after opening from a rubbish bin will usually yield no reliable information on the microbiological content of the canned product at the time of opening the can. However, detection of pathogens in these circumstances, such as C *botulinum* and its toxin or salmonella, may be significant in the early stages of an investigation.

Examination of samples in the food microbiology laboratory

Sample preparation

All samples, containers, plastic bags, bottles, test tubes, petri dishes etc must be labelled so they can be easily identified during all stages of the analysis to eliminate any possibility of confusion.[2]

Samples should be handled carefully to avoid risk of contamination. In the case of packed products, the outside of the packing should be decontaminated with an appropriate agent, eg 70% (V/V) ethanol. Any instrument used for opening the packing unit or which comes into contact with the sample must be clean and sterile.[2]

All foods other than most liquids need some preparation before they are analysed. The commonest method of sample preparation is to produce a homogeneous suspension of the food sample dispersed in an appropriate diluent. This is designed to give a uniform distribution of the microbes present in an easily pipettable form. For convenience, analysts traditionally prepare a 1 in 10 suspension to simplify any subsequent concentration calculations.

Quality control in the microbiology laboratory

Most techniques rely upon dispersing the food as small particles, although some techniques (eg orbital shaking or mixing) may rely on dislodging the microbes from the food. A common method is stomaching, or using a top or bottom drive macerator. The advantage of the stomacher is its ability to homogenize the food while it is sealed in a bag, reducing the risk of sample contamination and the creation of aerosols. Maceration techniques require sterilizable, sealed maceration units. Some dry samples have a period of rehydration before stomaching, in order to minimize the effects of osmotic shock to the microbes. Control of rehydration time and temperatures is important as microbial growth may otherwise occur (typical times at room temperature are between 30 minutes and 2 hours).[6]

The carrier liquid used should be sterile to avoid sample contamination and is designed to minimize damage or shock to any microbes present in the sample. Some carrier liquids have been developed for use with specific types of foods, eg fatty foods, which are difficult to homogenize.

Some samples require dilution prior to analysis in order to reduce the concentrations of microbes present to a measurable level. It is common practice to carry out serial or decimal dilutions, ie 1 in 10 dilutions. It is essential that the dilutions are carried out aseptically in containers capable of closure. All containers must be appropriately labelled.

The exact method of sample preparation, including the weights and volumes used, should be specified in the methods manual. Prepared samples and their dilutions should be analysed immediately and delays greater than 30 minutes recorded. If any further manipulations are needed (eg pasteurization for spore counts), then these should be specified and recorded.

Sample analysis

As noted previously, all methods should be clearly documented and referenced in the methods manual. They should have been evaluated and shown to be suitable for examining foods and related samples. Where possible, methods approved by the International Standards Organization (ISO) for food analysis should be adopted.

Many food microbiology techniques require the analyst to interpret as well as recognize and describe the results seen. Furthermore, many of the methods involve a series of stages. The final conclusion reached from the analysis depends upon *all* the stages being correctly performed, interpreted and acted upon. Therefore, it is essential that the analyst is numerate, properly trained and is regularly assessed for competence with reference to results produced using the documented methods on samples of known microbial content. In many cases the important stages are manual (eg diluting) and the analyst can have an important influence on the overall accuracy of the results.

It should be recognized that microbiological analyses of food have poor accuracy and reproducibility because of the heterogeneous distribution of microbes in many foods and inherent inaccuracies of most microbiological analytical techniques.[7] These shortcomings must be taken into account in analysis of the results.

Recording and interpretation of results

The analyst must record all the data obtained before conversion to a concentration measurement. This allows other analysts to check and confirm the results calculation. Some results are reported in a trans-formed state (eg present or not found in a measured amount of food).

The coding of results must be related to the coding of the original sample. Ideally, this code should be identical and simple to read. Unfortunately, this is not always possible, eg where pooled samples are used for analysis.

It is often left to the analyst to interpret the significance of the results obtained. This should only be done in consultation with an experienced food microbiologist, as the levels of various groups of microbes in a food, the expected types and the relationships of their concentrations can vary depending upon the components of the food, the process given and treatment since manufacture. For example, some pâtés are pasteurized in their packs and should contain only a few hundred sporing bacteria/g, whereas other pâtés are not, may be decorated or sliced after cooking and thus may be contaminated with a wide range of microbes, particularly lactic acid bacteria. A good analyst would be able to identify the type of pâté by interpretation of the data obtained.

Follow up analysis

All remaining portions of the food samples and any isolates should be held appropriately (eg at chill or frozen) until all the analyses and reports are complete. This ensures that retesting can be carried out if necessary.

The practice of splitting samples on receipt can be very useful in cases of further investigation or dispute over results.

Analytical reagents

To improve the reliability of results it is recommended that, for the preparation of culture media, dehydrated basic components or complete dehydrated medium be used. The manufacturers' instructions should always be followed. Basic components should always be kept in a cool dry place, sheltered from light, closed tightly and date stamped to assist stock rotation. Any chemicals used for the preparation of media should be of recognized analytical quality.[2]

Each batch of prepared media should be tested for selectivity using appropriate strains of a pure culture obtained from a recognized culture collection. The ability of a strain to grow on an appropriate medium acts as a quality control procedure for media preparation.[1]

References

1   Betts RP, Bankes P. A code of practice for microbiology laboratories handling food samples. Technical manual no. 21. Chipping Campden, Glos: Campden Food and Drink Research Association, 1989.
2   Anon. Microbiological examination of food and animal feeding stuffs. BS 5763: Part O. General laboratory practices. London: British Standards Institution, 1986.
3   Anon. Categorisation of pathogens according to hazard and categories of containment. Advisory Commitee on Dangerous Pathogens. London: HMSO, 1984.
4   Anon. The Health and Safety at Work etc. Act 1974. London: HMSO, 1974.
5   Anon. Control of substances hazardous to health regulations (COSHH). London: HMSO, 1988.
6   Anon. Microbiological examination for dairy purposes. BS 4285. Section 1.1. Sampling and preparation of sample. London: British Standards Institution, 1984.
7   ICMSF. Micro-organisms in foods, 2. Sampling for microbiological analysis; principles and specific applications, 2nd edn. Toronto: University of Toronto Press, 1986.
8   Cochran WG. Sampling techniques, 3rd edn. New York: Wiley, 1977.